The Phone Fix

DR FAYE BEGETI is a practising neurology doctor and neuroscientist at Oxford University Hospitals. She completed her medical degree and PhD at the University of Cambridge, and has since continued her research into neurodegenerative diseases alongside seeing her neurology patients. Her Instagram account @the_brain_doctor was started to share her knowledge more widely and has since amassed a community of over 135K. She lives in Oxfordshire with her husband and two young daughters.

DR FAYE BEGETI

The

Phone

Fix

HEAD
of ZEUS

An Apollo Book

First published in the UK in 2024 by Head of Zeus Ltd,
part of Bloomsbury Publishing Plc

Copyright © Dr Faye Begeti, 2024

The moral right of Dr Faye Begeti to be identified
as the author of this work has been asserted in accordance with
the Copyright, Designs and Patents Act of 1988.

9 7 5 3 1 2 4 6 8

A catalogue record for this book is available from the British Library.

ISBN (HB): 9781803285566
ISBN (XTPB): 9781035903009
ISBN (E): 9781803285542

Illustrations by Luna Aït-Oumeghar
Typeset by Ed Pickford

Printed and bound in Great Britain by
CPI Group (UK) Ltd, Croydon CR0 4YY

Head of Zeus Ltd
First Floor East
5–8 Hardwick Street
London EC1R 4RG

WWW.HEADOFZEUS.COM

For Steve, Lyra and Aria

*Sitting on your shoulders is the most complicated
object in the known universe*

Michio Kaku, theoretical physicist

Contents

Introduction

The Phone Age

Humans have always been worried that the world would be ruled by machines. Depicted in movies as malevolent and creepy, with uncontrollable powers ready to take over the planet, there is one key difference that those movies got wrong – machines in real life are a lot smaller and they have already taken over. Their inconspicuous nature has given them many advantages and, acting under the guise of being useful, they seem non-threatening and unobtrusive. There's probably one by your side.

We are bombarded with images comparing technology and our phones to drug paraphernalia, with headlines insisting that they are melting our intellect, or corrupting our children. One could be forgiven for thinking the apocalypse is upon us. But to be honest, the reality does not feel like that. Living with our robot overlords is not quite the scary hellscape I was expecting it to be. I quite like the ability to work out where I'm going, to easily keep in touch with my friends, to access a wealth of useful information.

Used in the right way, our phones make our lives better, but used in the wrong way, our inconspicuous devices can have an entirely different purpose. Rather than helping us achieve our goals, they can be a source of distraction from them, draining our time and impeding our focus.

I want you to think about this for a moment.

- Do you feel that you cannot control how much time you spend on your phone?

1

- Do you sometimes automatically reach for it without thinking?

- Do you often check your phone so frequently that not enough time has reasonably passed for something new to have happened?

- Do you get a strong urge to check your phone in particular situations, such as first thing in the morning?

- Do you look up from scrolling to find that more time has passed than you realized?

- Do you procrastinate starting tasks until you've caught up with all your apps?

- Do you sometimes wish you could put your phone away to focus on a meaningful task?

- Does your phone use interfere with your sleep?

- Have you attempted a 'digital detox', only to fall back into your old habits?

- Are you worried about the effect your phone will have on your brain?

If you've answered yes to any of these questions, then this book is for you.

How we got here

The Phone Age began in 2007 with the release of the first iPhone. I was nearing the end of my first year of medical school at the University of Cambridge and I should have been studying for my exams – except I wasn't. Eager for something else to do besides pore over reams of text, I walked around the beautiful library building in Jesus College and scanned the shelves. An intriguingly titled book stood out among the sea of textbooks:

The Man Who Mistook His Wife for a Hat. Authored by the eminent neurologist Oliver Sacks, this riveting work weaves together the tales of his patients' extraordinary and unusual experiences, showing readers what can happen when the human brain doesn't work as expected.

I picked it up and started reading.

I grew up at a time when the Internet was still dial-up and using floppy disks to store files was cutting edge. I needed to carry spare change to call my mum from the phone box to tell her I'd be home late. I would buy songs on cassette tapes but, when I couldn't afford them, I would use a tape recorder to record them straight off the radio. And when that radio stopped working, I would spend hours of my time taking it apart. Starting off with a goal of trying (and spectacularly failing) to fix it, I soon realized that trying to figure out how it works was just as much fun. Reading that book was the moment that I knew what I wanted to do. I had found my radio as an adult, and I couldn't wait to get started.

It turns out that the brain is a much more complicated puzzle to solve than my basic childhood radio. The brain is made up of nerve cells, known as neurons, the number of which is in the same order of magnitude as the stars in the galaxy. Despite their tiny size (most are less than a fiftieth of a millimetre in diameter), each neuron contains its own extraordinarily complex machinery. And, unlike the static manufactured parts of my radio, the nerve cells inside our head are living and in a constant state of change, our experiences causing these connections to strengthen and weaken in real time.

This was not a puzzle that I would be able to work on in my spare time. I therefore made a bold decision to interrupt my medical studies and undertake neuroscience research. My journey led me to the Brain Repair Centre, an aptly named building for someone who is trying to fix things. I spent three years in this ambitiously named department, running experiments to figure out which parts of our brain we use to think, what happens when those parts are damaged by neurodegenerative diseases, and if there are any ways that we can fix them. With a PhD under my belt, I returned to complete my medical training. I am now a practising neurology doctor and spend my time seeing people whose brain is not functioning as it should. My patients might describe trouble moving parts of their body, their memory might be failing, or they may be having seizures. As each part of our brain has a multitude of functions, it is the neurologist's job, much like a detective, to assimilate all the clues to pinpoint the exact location of the abnormality.

I have dedicated my entire career to understanding the inner workings of the brain but, above all, my overarching goal is to improve people's brain health. The World Health Organization defines brain health as an optimal state of function across several domains that allows an individual 'to realize their full potential over the life course, irrespective of the presence of or absence of disorders.'[1] This means that you do not need to suffer from a disorder to benefit from understanding how the machinery inside your head works. It's not just about treating disease; it's about maximizing your abilities and fulfilling your greatest potential.

Meanwhile, over the past two decades, I watched the technological landscape evolve. Advances in technology have supported scientific advances in ways that would not have been thought possible a generation ago. They have also benefitted us personally. Instead of floppy disks and cassette tapes, there are smartphones and social media. And no one needs to carry around spare change for the phone box. Instead, it's not unusual for patients to use their smartphones to share images or videos of their symptoms during consultations. Wishing to move beyond the clinic room, to reach more people and share insights about improving brain health with a wider audience, I created an Instagram account (@the_brain_doctor) to teach people how the brain works. Rather than infrequently scheduled neurology appointments with a limited time and an abundance of information to cover, I felt more people could learn useful strategies straight from their phone.

When the opportunity arose to write a book about the brain, I wanted it to have the maximum possible impact and to positively influence brain health. After all, every single thing we do, every action we take, starts off as a bit of electricity in a part of our brain. So, I asked myself – what is the biggest factor shaping our lives today?

To answer the question, I thought about where we spend most of our time. Apart from sleep, 'screen time' forms the single biggest activity that we partake in collectively as a society, both in and outside of work. The average person spends around four hours online each day, three of them on their smartphone and the rest divided between computers and tablets.[2] This accounts for 25 per cent of our waking time, assuming that we sleep for eight hours, although this may be less given that how we use our devices also impacts our sleep. It wouldn't be surprising if for many, especially those drawn to this book, actual screen time might very well surpass these average figures.

This is not a judgemental statement. We should be free to spend time in whatever way we choose. I've embraced technology with every ounce of my being. Whereas once I had to go to the library to get a book, I can now download it in an instant. Better still, I can even contact the author by email or find them online. Posting on social media provided the building blocks I needed in order to write this book, teaching myself how to write for a general audience, rather than a scientific one. I shared free expertise with my followers and in turn they provided equally valuable feedback on what worked and what didn't.

Yet most of us also experience our phones as being the ultimate source of distraction. Chances are that, if you are like most humans, at some point during reading this book your attention will begin to wander and you will immediately reach for your phone. It may be a 'quick check' to look at the time or your messages but, before you know it, you may be quickly swallowed down an endless scroll hole, jumping from link to link, your original purpose long forgotten. Time is distorted when you enter this virtual world and when you look up you may be surprised at how much time has passed.

This is where the crucial problem lies: not time spent on technology – but *unintentional* time. A lot of the time spent on our devices is not by choice. We are having trouble controlling it.

If you feel like you have a problem with your phone, I don't want you to consider this a gloomy state of affairs but an opportunity to change. The constantly evolving nature of our brain means that yours will be different by the time you finish reading this book. What's more is that we are in control of that change. Humans have thoughts, feelings and patterns of behaviour but we also possess something else – the capacity to assess them, analyse them and change them – a process known as metacognition. But to be able to harness that power you need some knowledge of how your internal circuitry functions, to know what is actually happening when you are distracted by your little device.

Much has been written about the influence of technology and there is a lot of fear dominating the debate about its impact on us and our brains. However, when objectively assessing the scientific evidence, much of that fear is unfounded. In addition, such fear leads to a communal anxiety with very little practical value. After all, being constantly told something is harmful can affect our brain in a way that this becomes a self-fulfilling prophecy.

Tired of the same scaremongering rhetoric that is being presented, I think it is time for a fresh perspective. Like every great civilization-advancing tool, our phones come with some downsides, and we can develop bad habits in how we use them, but we must be grounded in reality and avoid the moral panic that is being pushed upon us. We need to move past the unhelpful narrative that technology is toxic, or that we are all afflicted by a phone addiction, and start discussing these matters using a more balanced approach. Rather than throw our devices away, we have to learn how to live with them.

After deciding that this is where my scientific and medical experience can have the most impact, I buckled down to work. I have spent the last few years researching the topic and reading as many scientific studies as I could. In this book, I will summarize what the science has to say and help you understand the structure inside your head a little better. I will not be scaremongering about phone use. I think it's OK to use technology to help us in our day-to-day lives, and as a source of enjoyment. I will show you how to use it in a way that maximizes the good while avoiding the bad, giving you the knowledge and the practical tools that you need to create a set of healthy digital habits.

A lot of people look back at the pre-smartphone, pre-social media era that I grew up in and reminisce. It was a simpler time. They wish they could go back. Having grown up in that era, I don't. If anything, I wish I had been able to watch a YouTube video on how to fix my radio. I love the access and opportunities that technology gives us, the connection with other people, the exchange of ideas, the ability to publish my work.

I'd rather stay in the Phone Age. It is much more fun.

And we need not fear our little machines. We just need good digital habits.

How to use this book

This book will explain how and why we have developed our individual phone habits, and what we can do to change them. It is structured in three parts as follows:

Part I – The Groundwork

To start off, I will discuss why we might feel that our phones are addictive even though the science says otherwise. Rather than acting on fear, you will gain knowledge of what is actually happening in your brain when you check your phone and why this action has become difficult to control. I will introduce the key parts that form our brain's machinery and discuss the highly coveted quality of willpower. This section is designed to provide you with a significant mindset shift, from feeling helpless to being empowered, finishing with some foundational strategies that will revolutionize the way you approach moments of difficulty.

Part II – The Habit Puzzle

In this section, we're going to delve into the intricate science behind habit formation. I'll introduce you to my unique methodology – 'The Habit Puzzle'. This concept illustrates the four fundamental elements necessary for establishing and embedding habits in your brain, with a particular focus on factors that have contributed to the development of our pervasive digital habits. This exploration concludes with a detailed, hands-on guide, offering you practical tools and strategies to definitively reprogram your digital habits.

Part III – Unlock Your Potential

Lastly, I delve into the specific domains of focus, sleep, mental health, and social media, examining their intricate relationship with phone use and the digital world in general. Drawing from the knowledge you've acquired in the first two sections, you'll be able to address and troubleshoot areas you may particularly struggle with. This will enable you to critically examine their association with technology and devise personalized strategies for improvement and, in doing so, unlock your full potential. The book concludes with a section dedicated to future-proofing your brain, guiding you on how to best preserve and enhance your cognitive health in the face of future digital challenges.

The primary focus of this book is on our relationship with our smartphones. While problematic habits often revolve around smartphones – our most personal and ubiquitous devices – the principles discussed can apply to any form of digital interaction.

Throughout this book, in addition to phone use, I use the terms 'online world', 'digital world' and 'virtual world' to offer a broader context to the digital habits beyond smartphones such as tablets, computers, and the plethora of apps and online platforms that they give us access to.

This book is evidence-based, and so I will draw on science and quote scientific studies throughout, but I want to make one thing clear. People may think of science as unquestionable, as the last word, the ultimate answer, but, having conducted extensive research studies myself, I can tell you that science is profoundly messy. Sometimes a study may show something, but further studies are not able to replicate it. This calls into question whether the initial finding was true or whether that finding was down to chance. Think about it this way – if you do an experiment where you flip a coin a hundred times, you will not get the same result each time. Just because there is an equal chance of getting heads or tails, does not mean you will get exactly fifty heads and fifty tails (in fact, the probability this will happen is less than 8 per cent). There is an element of chance to every study and, of course, there are many more complexities and interlinking factors to studying humans than flipping a coin.

Sometimes, research studies completely disagree and have directly opposing findings. Interpreting data is made even more complex when looking at features of study design – who was tested, how the testing was done and when the testing took place. And, of course, the interpretation of data is ever-changing as more evidence comes to light. So, instead of a definitive answer, results are more like a tangled piece of string that we, as scientists, are constantly trying to unravel. I am going to use all my medical and scientific knowledge to make sense of this for you by presenting the research findings in the most clear and concise way possible. I will try to explain the complexity of the processes going on inside your head, trying to be as accurate as possible without oversimplifying but, at the same time, I will avoid getting into minutiae and scientific debates which, while captivating for scientists, offer little practical value to our everyday lives.

Another thing to keep in mind when reading this book is that when exposed to new knowledge, it is not unusual for our brain to get some a-ha moments and a little burst of motivation. In order to capitalize on this, I have included practical sections throughout with tools and tips to help you create healthy digital habits. I cannot emphasize enough the importance of making changes as soon as possible, even as you're reading. Rewiring in the brain takes much longer than people realize (in the order of weeks to years) and starting early, rather than waiting for a future perfect moment, gives you a head start on this process before distractions and procrastination can start to derail you.

As a rule, you should start to make changes before you feel ready. Our brain fools us into thinking that our future self is akin to a superhero – that there will be a more opportune time to start. This happens to me every day when I leave the washing up until tomorrow, thinking that I will have more energy, when in fact constant experience has shown that I am just as busy and tired the next day. Reading this book and working on your phone habits is the best way to look out for your future self. It is a time and attention investment that will pay itself back several times over.

Our lives – including the ways that we use our phones – are unique, so it is not possible for every technique and situation to apply to every person. My clinical advice differs even when patients have the same diagnosis. Our brains are shaped by genetics, socioeconomic factors, and different life experiences so I do *not* subscribe to the 'this has worked for one person and therefore must work for everyone' train of thought. I did not create this book as a fixed, rigid plan where every step must be followed meticulously but rather as a toolkit of theory and practice, to be adapted to fit your own unique situation and lifestyle. Much like a physical toolbox at home, you can assess the situation and decide what tool or tools to use now, saving the rest for later. Just because you wouldn't use a hammer to fix a hole in your wall, doesn't mean you'll never need it. Save that hammer for another time and revisit this book often. The relationship with our devices is something that has to be managed long-term and I am giving you the toolbox for building good digital habits in the Phone Age.

PART I

The Groundwork

1

Is Your Phone Addictive?

Keira* was born the same year that I was, but our lives had taken different paths. I was her doctor, and she was my patient. With a deep-seated infection in her heart, she was amid a several-month-long stay in hospital and receiving multiple antibiotics round the clock. The end goal: open heart surgery. One of the many dangers of illicit drugs, such as heroin, is that they are not clean substances, either in the way they are manufactured or in the method by which they are injected. Bacteria, once they enter the bloodstream, have a predilection for settling in the small valvular leaflets of the heart. After each heartbeat, these valves close to prevent the backward flow of blood, thus maintaining the body's natural one-way circulatory system. With the brain receiving 25 per cent of the blood pumped directly by the heart, each pump can lead to a stream of these bacteria creating pockets of infections causing damaging strokes. Keira would have to replace three out of four valves in her heart with metallic ones if she was to live. But, most importantly, she would need to tackle her drug addiction.

* The case studies in this book are based on my clinical experiences with real patients. However, to respect privacy and maintain confidentiality, I've deliberately changed names and other identifying details. Any resemblance to actual individuals is purely coincidental.

Addiction ruins lives, your phone does not

Our brain contains a large number of neurons, an estimated 86 billion according to the scientists who went to the trouble of counting them.[3] The best way to think about each neuron is like a tree. It has a trunk, the scientific term for which is the axon, and the end sprouts out into large branches which get progressively smaller. The ends of the smallest branches are called the dendrites, a word originating from the Greek word 'dendron', for tree. The dendrites of each neuron make connections with neighbouring neurons of which there are countless possibilities. When two dendrites oppose each other and make a connection, there is a small gap termed the synapse. To transmit an electrical signal from one neuron to another, chemical messengers called neurotransmitters are released into this gap and are then detected by the receiving neuron.

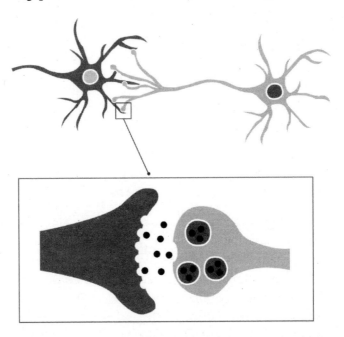

Our Neuronal Machinery: Our brain contains an estimated 86 billion neurons. The connection between two neurons is the synapse and signals are passed between them using chemicals known as neurotransmitters.

The majority of addictions are to substances that powerfully affect our brain. Irrespective of how they are consumed, all addictive drugs enter the bloodstream. They then cross the blood-brain barrier – a border designed to shield and protect our nervous system from potential threats – producing a potent chemical effect. They can do this in a variety of ways: they may mimic our chemical messenger molecules, stimulating a neuron to dump a large number of neurotransmitters into a synapse; or they might prevent the receiving neuron from reabsorbing the neurotransmitters and cleaning up the synapse. No matter the mechanism, one thing is the same. The chemical effect of addictive drugs is so powerful that it supersedes our own biology. Synapses are flooded with neurotransmitters, resulting in levels that are much higher than anything we would experience normally.

Addiction is a clinical disorder with stringent criteria. Colloquially, the term takes on a different meaning. We say that we are addicted to our phones, social media, or the latest binge-worthy series. You don't need to go far to find numerous articles describing the perils of 'phone addiction' and the supposed signs that help you determine whether you might be unlucky enough to be afflicted. Seeing these headlines or hearing these phrases often makes me think of Keira's turbulent stay in hospital, her multiple admissions to intensive care, and her eventual discharge with three pristine metallic valves.

Soon after Keira's discharge from hospital, I stumbled upon an article published in a leading newspaper. It was titled, 'Constant craving: how digital media turned us all into dopamine addicts', and likened the smartphones in our pockets to the 'modern-day hypodermic needle'.[4] This type of narrative is misleading and serves only to sensationalize an object that is an unavoidable and essential part of many people's lives. We cannot compare using smartphones and their social media and gaming apps to substance abuse and drug paraphernalia. The claimed 'dopamine hit' from phone use is not equivalent to the effect that addictive drugs have on our neurotransmitters; a difference likened by the National Institute on Drug Abuse to a gentle whisper in your ear versus someone shouting through a microphone.[5] Picking up your phone one too many times is not the same as being addicted to a substance. In addition to being scientifically untrue, it also minimizes the struggles that people with addiction face.

I saw Keira a few months later. Her newly replaced metallic valves were now full of bacteria, only now the infection was near impossible to treat. Synthetic materials in the body provide a shiny surface for bacteria to cling to but have no immune cells to defend against infection. The risk of reinfection in prosthetic valves is very high and something patients are extensively counselled about prior to surgery. The only solution, to keep their metallic valves clean, is to never inject drugs again. Keira had relapsed and that admission was to be her last.

One of the key criteria of diagnosing addiction is continued use despite severe negative consequences. Keira's addiction was so severe that, despite being in her prime, she underwent several surgeries before eventually losing her life. I've personally witnessed the effects of addiction, not only in patients like Keira suffering from infective endocarditis, but also during my time in the gastroenterology ward. We would have to remove the alcoholic hand gel that we used to sterilize our hands from the vicinity because patients, despite being admitted with life-threatening liver damage, could not help but ingest it as a substitute for alcohol.

It is possible to suffer from a non-substance addiction – termed a behavioural addiction – of which gambling was the first to be included in medical classification systems. These do not have the same effect on the brain's neurochemical system but they must have the equivalent severe negative consequences to be diagnosed. It is not uncommon for people with a gambling addiction to lose their jobs, destroy their relationships, squander life savings and get into severe debt. Deception is a big part of addiction too, with many addicts going to great lengths to hide what they are doing from their loved ones, potentially even committing crimes in the process. In essence, addiction ruins lives. To be addicted is to say that there is a degree of functional impairment: your life has to be severely impacted to a point that you cannot function. And smartphones, or the apps within, do not have negative consequences to the degree required to constitute an addiction. In fact, much of the time, rather than leading to functional impairment, they provide tools to help us function better.

So, what is driving this addiction label? The answer is moral panic.

A moral panic

A moral panic is a collective feeling of fear of an entity that is perceived to threaten the values or well-being of our society. Moral panics are often irrational. They happen because, simply put, our brain fears change. This is an evolutionary mechanism to help keep us protected. It means we are naturally suspicious of any emerging technology that we are not used to, much like we are cautious of people we do not know.

Growing up, I would listen to my grandmother tell me stories of how her family residing in a small village in rural Greece obtained a clunky mahogany-covered box that her grandfather thought would be the source of many evils. It was a radio. Turning the dials introduced her to a whole new world of music, news and entertaining shows that she looked forward to listening to. A century ago, their radio was viewed as cutting edge, as smartphones are today. It was also regarded with the same apprehension, particularly from the older people in the village, many of them outspoken about their suspicions. This new 'toy' would make the young people listening to it lazy, they said. Others thought it would be the ultimate source of corruption, both statements very reminiscent of the narrative currently aimed at smartphones. But in her nineties, my grandmother, with sheer delight on her face, would recall how she stayed up late to dance to music, all the while laughing at the old men's foolish notions.

Moral panics are not new. They can be traced as far back as Ancient Greece, where early philosophers such as Socrates feared that the ability to write things down on paper would ruin our memory. In the 1800s, the invention of the telephone was met with the same fear; in the 1920s, several headlines denounced the evils of the crossword puzzle; in the 1930s, it was motion pictures, followed by comic books in the 1950s and then video games in the 1970s.[6] The media's primary job is to sell papers (although they have had to adapt this somewhat given the rise of technology) and capitalizing on our natural suspicions with sensationalist headlines produces those sales.

What lessons have we learned from the suspicion that surrounded video games over the last few decades? In 2018, the World Health Organization officially recognized 'Gaming Disorder' in the International Classification of Diseases (ICD-11), despite its existence remaining a subject of ongoing debate among researchers.[7,8] The Diagnostic and Statistical Manual of Mental Disorders (DSM-5) includes 'Internet Gaming Disorder' under 'Conditions for Further Study', indicating a need for more investigation in this area. Conversely, 'smartphone addiction' was intentionally excluded due to a lack of substantial evidence from scientific studies supporting its existence.[9] As a result, doctors are unable to diagnose 'smartphone addiction' since it's not an officially recognized disorder. However, it is possible to diagnose 'Gaming Disorder'. But does this mean we are facing a widespread epidemic of video game addicts?

The short answer is no. Internet gaming disorder provides a diagnosis for a small minority of people who will completely forgo sleep and food to play video games and may suffer medical consequences as a result. In an extensive study of over 18,000 people, this was shown to affect less than 1 per cent of people who played video games. The majority fulfilled none of the criteria and suffered no ill effects as a result.[10] Despite many scaremongering headlines, we have not turned into a nation of addicts. In fact, it turns out that someone who plays video games is much less likely to become addicted to them than someone who drinks alcohol. Instead, more recent research shows that playing video games can enhance well-being by providing a psychological escape.[11] This is not surprising given that a common feature of a moral panic is that effects are widely overinflated, led by fear rather than evidence.

While there is little evidence of a rise in problematic addictions relating to technology, this doesn't stop judgement, more commonly from members of older generations who think that engaging with those activities is a waste of time. The same judgement that my grandmother faced for listening to the radio is very similar to the criticism that young people may receive today for interacting with their friends on social media. However, it is important not to equate 'waste of time' with danger. How people choose to live their life is ultimately up to them. Spending hours training hard and suffering negative consequences for the purpose of sporting greatness is considered commitment. On the other hand, using smartphones or playing video games for the purposes of fun goes against our celebrated notion of productivity. Calling something a waste of time is a societal judgement, not a scientific one. We are preaching an idealized way of living life when the whole purpose of autonomy is that we should all be free to choose.

This type of thinking has consequences, something I am acutely aware of through my medical practice. Making a diagnosis in the medical field prompts people to think that they have arrived at the correct answer and, as a result, they will stop trying to assess it critically. This type of complacency reduces analytical thinking and is commonly referred to as a 'premature closure bias'. This is where automatically assuming smartphones are damaging or addictive does more harm and supports a superficial way of thinking which fails to get to the root of the issue. People may be using their smartphones as a coping mechanism. They might find playing a game distracts from a negative mood or that scrolling on their phone helps them cope with social anxiety. Just like pain is the body's response to an underlying issue, smartphone overuse is often a symptom rather than a diagnosis.

The way we think about smartphones can be quite judgemental. We tend to place non-technological activities on a pedestal and look down on people who prefer to communicate through their phones. Technology provides a powerful communication tool and some people, such as those with mental health difficulties or neurodiverse brains, may rely on it to a greater extent. Pathologizing everyday behaviours that many people find helpful only worsens the stigma.[12] Blaming the smartphone shifts the responsibility away from deeper societal problems. We absolve societal expectations of constant productivity that make people check their email at night and blame the technology that enables us to do so instead. We condemn social media as the cause of mental health problems, not realizing that this is a symptom of something bigger, a society where perfectionism is desired and constant comparison is craved.

Technology habits

You may feel guilty about the amount of time you spend on your phone and have tried and failed to stop checking it on a number of occasions. You may also blame yourself for not having enough willpower to do so. This is because the phrase 'put your phone down' sounds beguilingly simple but masks the sheer complexity of our underlying neurology.

The brain has a powerful ability to change in response to our actions. 'Neurons that fire together, wire together', famously said by neuropsychologist Donald Hebb, has become a powerful maxim in the neuroscience world. 'Practice makes perfect' is another phrase which may be more familiar. Practising a musical instrument means your brain gets better at it. Practising picking up your phone does exactly the same. The reason you are unable to stop checking your phone is the same reason that someone may not be able to stop biting their nails or that someone else snoozes their alarm clock every morning. They are all habits.

Actions that we repeatedly perform get coded into our brain in a way that makes them easier. They become second nature, and herein lies the problem. Depending on what technological habits your brain has stored, your phone use can be a help or it can be a hindrance.

Technology habits are not unique. We develop habits in all areas of our lives and relating to an innumerable number of objects. However, what is unique when it comes to our phones is the abundance of habits that we have developed around a single object, reflecting their multi-functionality and close proximity. It is only possible to habitually snooze our alarm clock in the morning and in our bed; our phones, however, have infiltrated all aspects of our everyday lives and, as a result, are by our side almost constantly. The average person will pick up their phone approximately ninety-six times per day (or every ten minutes)[13] – a number that reflects both the amount of phone habits we have and how easily new ones become ingrained.

How habits are encoded in the brain means that we do them in a rather automatic fashion, in a way that overrides our original intentions. This automatic nature – for example, picking up your phone without thinking – can make those with strong digital habits feel as though they are addicted; however, this is not true in any medical sense of the word. Someone may resolve that they will get out of bed as soon as their alarm clock rings only to unconsciously hit the snooze button, but they do not suffer from an alarm-clock-snoozing addiction, just a problematic habit. Constantly getting distracted may reduce academic performance, procrastinating work may lead to missed deadlines and staying up late to binge-watch content may lead to tiredness but they are not enough to constitute a phone addiction.[14]

A knee-jerk reaction to this moral panic surrounding phone use is to measure screen time – to fret about it and then try to set arbitrary limits. Looking at the time we spend on our devices or on apps may provide some kind of initial wake-up call – it may even have led you to purchase this book – but scientifically the concept of screen time is flawed. There is no scientific evidence that there is a sharp cut-off point after which an amount of screen time becomes harmful. The time we spend on devices consists of a range of activities and compiling it into a single measure is far too crude and imprecise to have any real meaning. It is the equivalent of a patient handing me a single number comprising the sum total dose of all the medication they were taking, without the medication names. Focusing on a single value lacks information about content and pattern of use and does not give insight into our habits.

Work, communication, and leisure can all increase screen time in ways that are not problematic. My screen time increases every time I do a workout using a fitness app, navigate using maps or call my friends; all activities that are central to my well-being. Even when using an app for recreational purposes, such as social media for example, a thirty-minute scroll during a break is not the same as thirty brief, minute-long checks interrupting an activity you really need to focus on. It is not unusual for someone with problematic phone habits to engage in near-constant checking of their phone. These quick checks are much more disruptive to our attention than longer, infrequent checks, even if they add up to the same amount of time.

It's not about time but about habits; whether you use technology intentionally or unintentionally, whether you use it as a tool or a distraction. The type of content we access is also a factor because time spent is a measure of quantity but conveys very little about quality. This is why a group of over eighty-one scientists from around the world wrote an open letter to the UK government to vigorously campaign against setting arbitrary screen time limits based on hype[15] and why I won't be setting any in this book. I will instead guide you on how to build digital habits to use technology in a way that best fulfils your individual purpose and well-being.

As a society we are becoming increasingly preoccupied by using measures to determine our health-tracking, heart rate, minutes spent exercising and hours slept. It is important to keep in mind that these measures are just proxies for our overarching goals; they are not the goals themselves. An obsession with measures can also have a potentially negative impact and lead to disordered behaviour. For example, the term orthosomnia describes a state when the pursuit of healthy sleep becomes an obsession.[16] A patient might over-rely on sleep trackers and, when their data is less than perfect, they became so anxious that, contrary to their original goal, their sleep is paradoxically worsened. Catastrophic thinking can become a self-fulfilling prophecy.

With the anxiety surrounding phone use, I think it's important not to develop a new complex surrounding screen time. It is reasonable to measure your screen time as a starting point but it can be detrimental to go to extreme lengths to meet a specific target. Having a balanced approach means when a screen time monitor pops up on your phone to tell you that screen time has increased, you shouldn't be worried. And, if a loved one messages you on a day where you've already spent several hours on your phone and you want to answer, you should. I certainly would.

A new mindset

I included Keira's story as a reality check, so you can appreciate what addiction truly is. Overcoming addiction is hard and professional help is almost always required. In contrast, changing your habits does not require professional help. Reading a book wouldn't have been enough to help Keira but it can help you. In the scientific world, terminology is starting to shift; more scientists are favouring terms such as 'problematic use' or 'habits', and this is the language I will use in this book, reserving the term 'addiction' for disorders whose severity requires it.

Instead of viewing our phones as a drug, let's consider them in the same way that we do our favourite food. For me, a chocolate croissant is a delightful treat, but if I was to replace every single meal with it, it would become problematic. I could resolve never to eat a chocolate croissant ever again, but this would be excessive given this is something that I enjoy. A better approach would be to include them in my diet in a balanced way. Embracing things that you enjoy does your brain a whole lot of good and, if I live to be as old as my grandmother, I will think fondly of all the times I was able to enjoy a lazy morning texting my friends with a cup of coffee and a freshly baked croissant, rather than disparaging myself for not eating more porridge and getting straight down to work.

This is how I want you to think of your phone. It's time for a new mindset. It is OK to use and to find aspects of your phone entertaining and enjoyable but it is also important to be mindful that your digital diet does not overwhelm everything else in your life. To find out how to do that, you must learn about how the brain works. I would therefore like to introduce you to two main operatives that reside inside your head: the executive and the autopilot.

2

The Executive and the Autopilot

Everything you have ever done, every thought you have ever had, every feeling you have ever felt started off as a little spark of electricity inside your head. This spark is called an action potential, a signal that is transmitted along the length of our nerves at 119 metres per second; the equivalent to over 260 miles per hour, a speed that exceeds that of the fastest Formula 1 car. Just like the engine of a supercar, to achieve its function, our brain consists of several key components. Each component has its own specialized role but all of them work together to make us who we are. Our entire life, personality and skills are stored in a structure weighing around 3 pounds – and that includes whether you have a habit of reaching for your phone.

Inside our brain, there are two key systems that govern our actions. The first is the executive. The executive is the boss. Its headquarters are in a part of the brain called the prefrontal cortex which is found right behind our forehead. The actions the executive makes on a daily basis include focusing our attention, planning ahead, making decisions and regulating our emotions. Those actions are collectively referred to by scientists as 'executive function'. The executive is responsible for deciding our long-term goals and taking actions that align with them. We, as humans, have the most developed executive of all the mammalian species, something that makes us very future-oriented.

It may be the boss, but deliberation by the executive is a slow and inefficient process. Our lives are too complex for the executive to be able to contemplate every single decision. So, what the executive does is to delegate, and it does so by allocating a proportion of tasks to another region. Found deep in the centre of our brain, there is a set of structures that are collectively known as the basal ganglia – this is our autopilot. The autopilot does not have any deliberation capabilities and instead stores a set of pre-programmed sequences which, when activated, produce quick and efficient actions. Those sequences are what we know as habits.[17]

Every habit we have acquired within our lives initially started off as a deliberate choice by the executive. This initial choice may have been decades ago, such as the time of day we have a shower or when we brush our teeth. Other habits may be more recent, such as checking email as soon as we sit at our desk or cycling through our social media accounts while still in bed. After a while, when these behaviours have been repeated enough times, we no longer need to make that choice. They become pre-programmed

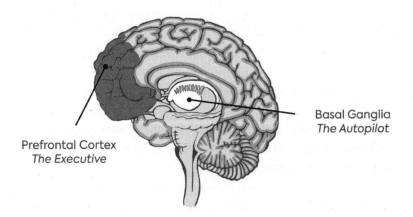

Basal Ganglia
The Autopilot

Prefrontal Cortex
The Executive

The Executive and the Autopilot: The two main operatives that govern our actions are the executive, in the prefrontal cortex, responsible for sustaining attention, planning, emotion regulation and long-term decision-making, and the autopilot, based in the basal ganglia, responsible for performing our habits but which only has the capability to make short-term decisions.

sequences to be stored in our autopilot and implemented without input from the executive. This is how our biology works to make the running of our brain more efficient.

Habits proceed so smoothly and efficiently that we hardly notice them. When you wake up each day, it is likely that a number of decisions will have already been made for you: whether you snooze your alarm clock, get out of bed straight away or reach for your phone all depend on the pre-programmed sequences stored in your autopilot. When travelling along a familiar route or doing an everyday task, your autopilot will take over, leaving the executive to do what it does best: think. In fact, this is a key feature of many habits – our mind is left to wander. During many autopilot tasks, our forward-thinking executive will use that time to think ahead and plan. Alternatively, we may think about the past, ruminating on something that has already happened and trying to process our complex emotions.

Automaticity is a key defining quality of habits and, for a quick real-life test to check whether something is a habit, and therefore can be done in autopilot mode, ask yourself the following questions:

- Is this an action that you are able to do with little concentration?
- Are you able let your mind wander and think about something else besides what you are doing?

Examples include getting dressed, cooking, cleaning, driving, walking, eating a sandwich, drinking a cup of coffee and, of course, picking up your phone.

Autopilot mode

We spend a large proportion of our lives in autopilot mode. In a series of influential studies, Professor Wendy Wood, an expert who has spent over thirty years studying habits, showed that approximately 40 per cent of our daily actions rely on habits.[18] This figure may come as a surprise to those who do not consider themselves to be routine-oriented but, rather than some of us being creatures of habit and others being free-flowing spirits, we all rely on our autopilot mode to a similar extent. Neither age nor personality seems to affect this. This just highlights how the automatic and subconscious nature of our habits means we often overlook the multitude of small decisions that we did not consciously make each day.

The efficiency of not having to consciously think about every detail of our daily routine helps conserve our mental energy. Starting a new job or school can be mentally exhausting because our habits are temporarily disrupted. We have to rely on our executive to accomplish even the most basic tasks, such as navigating the physical layout of our workplace or learning the location of important tools and computer systems. As we settle into any new environment or routine, our brain develops a new set of habits that become stored in the autopilot. Being able to perform those basic tasks on autopilot mode, without requiring the full attention of the executive, then reduces our mental strain.

Anything that doesn't require our full concentration will either be a habit or have many habitual elements. For instance, cooking a new meal from scratch can be quite challenging the first time we do it because we are mainly dependent on our executive and have to painstakingly think about each step of the process. However, once we repeat the recipe a few times, the autopilot is able to step in and ease the load on the executive, and we can start to simultaneously do other things, such as chatting or listening to a podcast.

In the past two decades, our brains have learned new habits. We tap, double-tap, swipe, and hold our fingers to glass screens. For those new to this technology, mastering these actions can be a challenge. But once these movements are stored in our autopilot, using technology becomes intuitive and seemingly effortless. These ingrained habits allow us to navigate various systems and screens with ease. To really see your habits in action, try switching to an unfamiliar operating system. You'll find that when you cannot rely on your autopilot mode, everything requires more thought and effort.

With almost half of our behaviour shaped by habits, our auto-pilot mode can have a significant impact on our identity. An 'organized' person, for instance, has honed hundreds of habitual sequences to keep things tidy, while someone viewed as 'messy' hasn't developed these routines. Our habits also influence how we interact with others, for better or worse. Some individuals may habitually show kindness or listen empathetically, while others may have the habit of frequently checking their phones during conversations, a behaviour known as 'phubbing', giving the impression of being disengaged.

The subconscious nature of habits makes them difficult to notice, let alone remember them accurately. To gain insight into our smartphone usage, a group of researchers from the London School of Economics mounted a small, unobtrusive camera on a pair of glasses that participants were asked to wear – this enabled them to obtain a first-person perspective into the participants' world and track what so often goes unnoticed. The study found that, on average, the participants interacted with their phone once every five minutes, often picking up their phone at times that they did not intend to, or spending longer on it than they had planned.[19] When shown the video footage of themselves picking up their smartphones, participants were surprised at how automatic the action was and how little they remembered it. They described the action of reaching for their phone as feeling natural, automatic, or unconscious. One participant even likened it to 'putting your hand over your mouth when you cough', a perfect example of an action taught early in life that eventually becomes second nature.

Problematic habits

Broadly speaking, habits can be classified into three categories: supportive, contradictory and neutral. As the names suggest, supportive habits are those that align with our goals and help us achieve them, while contradictory habits conflict with our goals and may hinder our progress. For example, sitting down at our desk each morning is a habit that supports studying. Scrolling on our phone at that time is not. Neutral habits have little impact in either direction, such as whether we brush our teeth before or after having a shower or getting dressed in a particular order. Often, these are simply referred to as good and bad habits. Though I might occasionally use these terms for simplicity in this book, it's crucial to remember that habits shouldn't be a basis for moral judgement. Having certain contradictory or problematic habits does not make someone a bad person. If you ever feel overwhelmed by the number of habits you feel you need to change, it's important to remember that, due to their contradictory nature, 'bad' habits will always stick out more than the myriad of supportive or neutral habits we have. It is up to you, and only you, to decide which category a habit falls under based on your individual goals. What someone considers a contradictory habit, someone else might feel is neutral, or even supportive, given their circumstances.

It is not uncommon to have several habits that conflict with your goals, but how does this happen? It is due to a fundamental difference between our two systems. With the executive brain, we can think long-term and make plans for the future, and one of the unique qualities of humans is that we endure temporary discomfort and delay gratification for an ultimate goal. This could mean long study sessions to obtain a degree, working hard to advance our career, or pushing past pain to achieve athletic success. Anything that requires us to give up a short-term reward for a long-term one depends on our executive system. In contrast, our autopilot is more focused on the present moment and immediate rewards. Its main function is to execute our stored habits to conserve mental energy. 'Cutting corners' so to speak means not having to think and this saving of limited mental energy is inherently rewarding. However, it also means that, if we have problematic habits stored in our autopilot system, it will continue to apply them even if they conflict with the goals set by our executive brain. In situations where no habit is stored and the executive delegates decision-making to the autopilot, the autopilot may act impulsively and prioritize short-term rewards over long-term goals.

Let's consider some familiar examples of habits. Leaving dishes next to the dishwasher, rather than inside it, is a habit that conflicts with the aim of keeping a tidy home. Even though placing them inside requires minimal extra effort, we often opt for the slightly easier route, which merely postpones the task. Similarly, taking a few extra seconds to consistently place your keys in a specific spot can prevent frantic searches later, yet we often toss them in the most convenient location in the moment. Upon reflection, these habits don't seem logical, but our autopilot mode isn't built for careful reasoning. It merely executes our ingrained sequences, continuing to do so even if it means deferring something that will complicate things later.

Checking our phones isn't inherently a bad habit, as some checks are crucial. However, many phone habits embedded in our autopilot mode can contradict our overarching goals. For instance, compulsive email checking can derail our objective of focusing on important work, while mindlessly scrolling through social media late at night might interfere with getting enough sleep. We might find ourselves reaching for our phones to check notifications when not enough time has conceivably passed for anything significant to have happened. If we thought about these actions rationally, we probably wouldn't do them. But habits are automatic and bypass our conscious decision-making processes, which means we might persist with habits like an early-morning social media scroll even when it makes us late for work.

Our autopilot habits also shape our behaviour in a more subtle way by helping us navigate an environment laden with more information and choices than our executive brain can handle. In scenarios where absolute freedom of choice could be overwhelming, our habits automatically whittle down the options. We think we are making a free choice, but in reality, our habits have already prepared a shortlist. For instance, when we're making breakfast, our brain doesn't sift through every food item available in our cupboards. Instead, it selects from a shortlist of our most frequently eaten breakfast foods. This is why we might repeatedly overlook a well-intentioned purchase of health food until it's past its expiration date, defaulting to our routine snacks instead, or why tinned foods are often forgotten until they've expired in the back of our cupboards.

Deeply ingrained technology habits function similarly. They provide an automatic shortlist of how we choose to spend our time. When we arrive home after a long day, our pre-set routines significantly influence how we relax and unwind. Rather than exploring a multitude of options, we might automatically gravitate towards our devices, opting to mindlessly scroll or binge-watch, thereby neglecting other hobbies, passions, or skills we wish to nurture. Digital habits can have an even more potent impact if they interfere with vital health activities like exercise, sleep, social interaction or time spent in nature. Our lives are shaped by how we use our time and where we focus our attention. Therefore, it's crucial to consciously decide how much of these valuable resources we wish to allocate to technology, and to foster healthy digital habits that enable us to do just that.

We often recognize that some of our tech habits are unhelpful and decide it's time to use our phones less or 'switch off'. To accomplish this, we apply maximum resolve to enforce a new pattern but the constant effort required of the executive brain to maintain this is mentally exhausting. It inevitably, at some point, fails and we return to our default settings – the habits in our auto-pilot – and berate ourselves for it. It can feel as though we are losing control when we find it difficult to stop doing something that we want to, our habits being executed without our awareness. We conclude that it must be our fault. We lack willpower. This is exactly what we need to discuss in the next chapter.

3

The Executive Battery

It is not unusual for some of the greatest discoveries in neuroscience to have been accidental. This was the case in 1848 on an otherwise ordinary day in Vermont where a man by the name of Phineas was working as a construction supervisor, a highly regarded and hardworking one at that. He and his team were excavating a site in order to build a railway track, clearing larger rocks through means of controlled explosions. To do this they would drill a hole deep into each rock and fill it with explosives, which were then topped with sand and pressed down using a long iron rod. This process is called tamping and was important to render the explosions both effective and safe. Tightly packing the material deep into the hole meant that, when the fuse was eventually lit, maximum force would be applied along the base of the rock, rather than towards the surface.

The events of 13 September 1848, at approximately 4:30 pm, would be dissected in neuroscience journals for the next 150 years. It is possible that not enough sand was packed on top of the explosives. It could be that Phineas was distracted by something his men said but, in the process of trying to compress the explosive material, the iron bar he was using scraped along the rock, creating the tiniest spark. No matter the cause, the outcome was immediate – the explosives ignited. The 13 lb, 1-metre-long iron bar he was holding flew out of his hand at immeasurable speed and pierced his left cheek, penetrating his skull and passing through the part of the brain found behind his forehead. It eventually exited at the top of his head; such was the potency of the blast that the iron bar came to rest a staggering 30 metres away.

Phineas Gage was twenty-five years old when this happened. Reports differ regarding whether his men carried him to a nearby cart or if he was able to walk by himself. Remarkably, Gage's sense of humour remained intact, and he even joked with the physician who came to assess him. However, in the following days, he became ill with an infection so severe that no one thought he would survive. Nevertheless, he defied medical expectations and recovered with what was considered minimal loss of function, just reduced vision in his left eye and some facial paralysis. However, his executive brain had been permanently damaged. In a letter written several years after Phineas' death, Dr John Martyn Harlow, the physician who initially treated him, wrote that, following the accident, the once-hardworking Phineas became 'impatient of restraint or advice when it conflicts with his desires'. The conclusion that 'he was no longer Gage' has echoed throughout neuroscience history.[20]

You do not lack willpower

Prior to Phineas Gage's accident, the function of the brain itself, let alone its different components, was completely unknown. His condition helped us discover that our unique human qualities rely on the part of the brain found behind our forehead. One of the traits that make humans different from other mammals is our ability to exercise restraint and self-control. The executive is the source of our willpower and Phineas Gage, having lost a portion of it, no longer acted in the same way. While the effects that this injury had on Gage are thought to be overestimated, many reported after his death at a time when he could no longer be tested,[21] it is common for someone to become more impulsive when the executive is damaged through a traumatic brain injury or a medical condition, something termed dysexecutive syndrome.[22] However, even when that part of the brain is functioning well, most people I meet wish they had more willpower. We have all had instances where we feel unmotivated and, instead of pursuing a difficult option with long-term benefits, we choose the easy short-term option. We might incessantly check our phone instead of working or forgo a workout to binge-watch a new series. We do all this while berating ourselves for our perceived lack of willpower.

It is not uncommon for someone to say, 'I don't have any will-power,' but willpower is not something that you are either lucky enough to possess or that you're completely missing – it's something we all have as part of our executive function. This does not mean to say that our brains function identically. Even though we cannot see it, executive function differs between people and, as a result, willpower does too. In a similar manner to our height, most people cluster around an average mark with smaller proportions of people sitting at the extremes of a scale. Genetics may play a role in our natural tendencies but our environment and how we use our brain also contribute.

The executive battery

So, what causes us to feel this so-called 'lack of willpower'? Simply put, our willpower levels are not constant and are in flux. The best way to think of your executive brain is like a battery. If you have a fully charged battery, you get that oomph. You are ready to go and hit your goals. But the opposite is true when that battery is drained. In that fatigued state, you are no longer able to muster the mental energy needed to do the difficult task, defaulting to the easy option instead. In essence, even though levels of executive function may vary from person to person, executive fatigue, and therefore lack of willpower, is something we have all experienced.

Generally, any strain on the executive will drain the battery. A lot of this strain is inevitable and forms part of our daily lives. Most people will appreciate that physical states, such as being hungry or tired, can affect your brain and drain this battery but its depletion is not synonymous with physical tiredness. It is possible for your executive to become exhausted through cognitive fatigue without any physical exertion. An example of this is focus. To be able to focus, your executive brain must sustain attention while ignoring irrelevant distractions, which means exercising a degree of willpower. How depleting this is for the executive will differ between people and it is also task-dependent. Spending countless hours sitting at a desk means little to no physical exertion bar moving fingers on a keyboard, but a challenging task that requires attention, such as writing this book, will generate notable cognitive fatigue when compared to the same amount of time spent casually surfing the web. We often don't appreciate or acknowledge the cognitive resources deployed when managing complex and challenging tasks, failing to fully understand the strain this puts on our executive brain.

The emotional states that we face also affect our executive. This is because a key function of the executive is processing the automatic emotions generated from the emotional parts of the brain, rationalizing them and deciding whether to act on them or not. This process is called emotion regulation and also requires executive power. For example, having an emotionally charged day at work where you have to deal with a lot of stress but still have to remain professional, has a larger draining effect than an otherwise uneventful day, despite working the same hours.

If you're a parent, you face a particularly difficult challenge: you have to manage your own emotions as well as those of your children, who are just learning how to handle theirs. It is common to see children, my own included, becoming upset or angry and having tantrums over seemingly minor things. This is due to their immature executive struggling with emotion regulation. A parent's brain has to simultaneously regulate their own emotions, by trying to remain calm, while helping their child process their feelings. They are emotionally regulating twice and, as a result, using much more executive power. Having to juggle the various activities and decisions that come with having children adds to this depletion, along with the physical demands of parenthood.

It is my experience that the executive-draining effects of being a parent are underappreciated even by parents themselves.

While these situations may deplete our executive function, and hence our willpower, on a daily basis, there are methods that we can use to recharge it. The most effective means of doing so is sleep, which has powerful restorative effects on our brain. A meta-analysis, aggregating data from multiple studies totalling 54,670 participants, found that both sleep duration and sleep quality were linked to levels of self-control.[23] Interestingly, this effect was more pronounced in older individuals. This means, as we age, we become more sensitive to the effects of sleep deprivation, making the late nights we managed in our youth a distant memory. Additional studies emphasize that adequate, quality sleep bolsters our mental resilience.[24] Generally, after a good night's sleep, we awake with a refreshed mind and renewed willpower. This is why most of us find it easier to tackle difficult tasks at the beginning of the day.

Satisfying our physical need of hunger is also important for recharging our executive battery, and it is not uncommon to become irritable when hungry (or 'hangry' as it is commonly known). This irritability is a sign that our hunger has depleted our executive brain enough to affect its emotion regulation capabilities.

Taking regular breaks also gives our executive respite and prevents the battery from getting too low. A break does not necessarily mean doing nothing. What we do in our breaks and the effect this has on our executive is specific to each of us, but there are a few things that executive-energizing activities usually have in common. Firstly, we typically find these activities inherently enjoyable, which means we don't need to exert additional willpower to get started. We usually have a sense of control over their duration, as pushing ourselves beyond a certain point can be depleting. Secondly, these activities often offer a degree of escapism and require less processing of complex information. There may be an element of challenge but not to a degree that it becomes frustrating. Examples could be reading, painting, cooking and catching up with friends but also watching TV, browsing the Internet, playing games and scrolling on our phone. Both digital and non-digital activities fulfil the above criteria. The uniqueness of our brains means that it is possible for the same activity to have a directly opposing effect on an individual's executive. For one person, cooking might be a source of enjoyment, whereas for someone else it might be a chore. Likewise, some may find the escapism provided by social media refreshing while others might find processing the same digital information depleting.

These differing experiences can be attributed partly to our unique natural inclinations and partly to the fatigue of our individual brain networks through use. For someone with a natural inclination towards reading and solitude, a book may provide a refreshing escape after a day filled with social interaction. Conversely, someone who has spent a solitary day immersed in words but is a social creature will prefer to catch up with friends. You can think of replenishing your executive brain similarly to resting an overused muscle, although it's crucial to remember that unlike muscles, the brain never truly 'stops'. It is always active, so the key isn't complete rest but rather strategic engagement in your personally energizing activities.

Low power mode

As we go through our day encountering and dealing with different stressors, our executive battery will deplete. The rate of depletion depends on both the complexity of the task and any emotion regulation aspect as both demand substantial executive involvement. Moreover, the challenges that our executive faces have the potential to compound, increasing the rate at which our battery drains. At the same time, we may not be charging it enough, and an insufficient amount of sleep or poor sleep quality can mean we start each day with an even smaller supply of willpower than the last. The unremitting use of this battery with no breaks means that it gets perilously close to running out.

If our executive battery was the equivalent to the battery on our phone, at some point it would cease to work. Complete cessation of function of our executive would be a real threat to our survival, so our brain does something else instead. Noticing that fatigue is setting in, it switches tactics to conserve energy, entering a state that I will refer to throughout this book as 'low power mode'. I have deliberately chosen this analogy for its familiarity – we all know how our devices limit their functionality, operating at reduced capacity to save battery power in low power mode. When our brain enters its own version of low power mode, a similar downshift in performance occurs. The executive steps down and, as a result, our executive function deteriorates – this means that our attention, working memory, focus, and emotion regulation all suffer. Situations that would normally be manageable with a fully charged executive suddenly become difficult to handle.

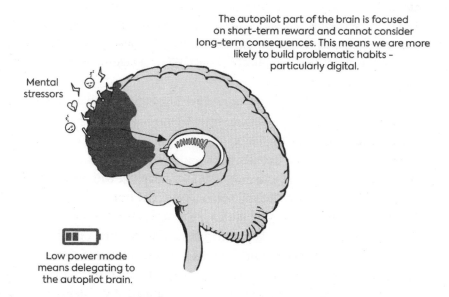

The autopilot part of the brain is focused on short-term reward and cannot consider long-term consequences. This means we are more likely to build problematic habits - particularly digital.

Mental stressors

Low power mode means delegating to the autopilot brain.

Low Power Mode: When our brain enters a 'low power mode' state, it becomes more short-term thinking. In this state, we are prone to developing problematic digital habits that then need more executive effort to overcome. This further depletes our executive reserves, making entering 'low power mode' even more likely, and so creates a vicious cycle.

When the executive brain enters low power mode, it delegates more decisions to the autopilot. As a more primitive brain region, the autopilot does not fatigue in the same way, but this energy efficiency comes at a cost. The autopilot's usual job is to implement our habits – repeating actions that we have done before – and in low power mode, it continues to do so but, upon encountering a situation where no habit sequence has been stored or has a range of behaviours to choose from, it will default to the easiest option, as it only considers the present moment. Low power mode converts our brain from long-term thinking to short-term thinking, where we are no longer willing to expend large amounts of effort for distant rewards, preferring to conserve our energy and obtain short-term rewards instead.[25]

It is important to note that our willpower has not run out. In low power mode, some willpower remains; it just becomes much harder to access. What is in reserve is important for our protection and can be summoned should the stakes be high enough. When in low power mode, you may not feel able to muster enough willpower to get off the couch, but in an emergency those reserves would kick in. In the past, an emergency may have involved running away from a wild animal but, in modern times, having to take someone to the hospital is more likely. High stakes could be a potential danger but equally they could be a large reward. After a draining day at work, you may forgo the future health benefits of a workout, your brain finding them abstract and difficult to conceptualize; but, for example, if a large enough monetary prize was on offer, you would be able tap into those reserves. Essentially, in low power mode, our executive brain has raised the threshold on what it will expend willpower for.

A notable feature of low power mode, and one which you may recognize, is that our ability to regulate our emotions effectively also dwindles. This diminished capacity can manifest itself in several ways. For example, our resilience towards trivial or minor situations may be considerably reduced. Insignificant circumstances or minor irritants can become sources of stress or frustration. An everyday situation like a delayed train, a minor disagreement, or even a simple mistake can provoke a more intense reaction than it usually would. For individuals who already grapple with a constant negative internal dialogue, this state of low power can exacerbate these feelings. The usual guardrails of reasoning that keep irrational thoughts at bay – 'not being good enough' being a common example – become less effective. Observing yourself closely, you may find that any self-criticism you have becomes louder and more pervasive after a particularly challenging day at work or during periods of sustained pressure or stress – this is something that affects everyone, neuroscientists included!

Scientists initially thought that short bursts of effortful activity, such as resisting a plate of cookies or solving some difficult puzzles, led to a reduction in willpower, but this has turned out not to be the case.[26] Unlike the fuel gauge of a car, which shows a reduction after each mile driven, executive fatigue sets in over hours rather than minutes.[25] These findings are in line with my own personal experience of testing patients' executive function. Despite patients' high degree of personal motivation in wanting to help advance research on incurable diseases, there is a progressive decline in how much energy they can muster in the second, third and fourth hours of testing, and their performance progressively drops.

For most people, being in low power mode reflects a more chronic state of overusing and undercharging. This was well illustrated in a series of studies comparing the brains of those who were assigned to do tasks of varying difficulty. In these studies, participants were presented with random individual letters on a screen and asked to recall whether the letter that was presented was the same or different to the one shown immediately before. For half of the participants however, the difficulty was significantly increased. Instead of comparing the current letter to the letter they had just seen, they were asked to remember whether it was the same as three letters before. After six hours of these depleting tasks, with a ten-minute break for lunch (the equivalent of a working day for many people reading this book), the participants were placed in an fMRI scanner. This scanning technique is able to measure changes in blood flow to different parts of the brain, allowing the researchers to see which parts of the brain are active. Those who had done the complex version of the task showed reduced activation of their executive.[25] It had stepped back and they were in low power mode. Further studies showed that being in low power mode impacts people's decision-making. When faced with the option of exerting minimal effort for an immediate reward or investing more effort for a much larger, delayed payoff, the fatigued participants were more inclined to choose the immediate, easy reward.[27]

Executive drain

Imagine you are nearing the end of what has been a long week. You have spent more hours at work than usual trying to meet a looming deadline. Ordinarily, you would be able to recharge somewhat during your commute by immersing yourself in an audiobook, but on this day, an unexpected bout of traffic and the resultant detours required extra mental effort by your executive brain. You get home later than expected, and an emotionally charged evening follows. In addition to work pressure, there are personal stressors that are playing on your mind. When the time comes to go to sleep, you realize that you've not stopped. Truly exhausted, you know that while you should close your eyes, you've not had a single second to yourself. You need to wind down. You pick up your phone. Scrolling feels like a welcome distraction from all the events of the day. You get to sleep much later. You are sleep deprived as a result.

Sitting down at your desk the next day, you feel tired already. Well and truly in low power mode, you cannot summon the energy to even start, so you get up to make a cup of coffee instead. You sit down again. You pick up your phone. No new messages. A quick glance at your inbox to see if there is anything urgent. You spend a few minutes browsing the web, followed by a quick check of social media and then a cursory glance at the news.

You've ended up in a scroll hole. Before you know it, a substantial amount of time has passed before you even start work.

Reaching for our devices is not uncommon in low power mode. It is an easy task that provides immediate rewards and many people use it as a method of psychological detachment, to switch off. We are able to browse a lot of information using our autopilot, and the action of so-called 'mindless scrolling' is actually an effort to give our executive brain some rest. Unfortunately, these actions form a template for future habits, and this is how digital habits slowly encroach on other parts of our life. The repeated action of reaching for our tech means that our brain begins to consider it a crucial part of our routine, like having a shower and brushing our teeth. Our brain, wishing to streamline these actions, will start to store phone-checking sequences in our autopilot. It is not long until every time you sit at your desk you automatically go through the same cycle without thinking. You have now developed your very own digital warm-up routine.

The nature of our habits can either preserve or drain our executive function. Society celebrates motivation and it is commonly assumed that successful people have more willpower. Superficially, it may seem that some people are able to avoid short-term distracting rewards, such as checking an entertaining app, by exerting their willpower. This is not the case and studies actually find that they use this precious resource less due to their habits.[28] Having supportive habits in our autopilot reduces the strain on the executive and can safeguard our willpower from becoming depleted. Remember that, even when fully charged, the executive will still delegate a proportion of tasks to the autopilot, this proportion rising sharply in low power mode. Developing supportive habits that can safeguard our willpower supplies is a key strategy to avoid entering low power mode.

Contradictory habits that conflict with our goals are, on the other hand, a constant drain for our executive brain. For example, if someone has a deeply ingrained habit of checking social media at their desk, they will have to exert willpower to stop. The continual effort of resisting that habit will drain willpower supplies, making them more likely to enter low power mode and develop further problematic habits. Although we form these negative habits in low power mode, once stored, they continue to drain our willpower even when we are no longer mentally fatigued. This extra pressure on our executive means we are likely to enter low power mode yet again and, as a result, develop even more contradictory habits, meaning we get stuck in a self-perpetuating cycle.*

Maximizing willpower

When it comes to willpower and motivation, it is first important to get the basics right. Good sleep, nutrition and taking adequate restorative breaks are important for the executive brain to function optimally and for us to maximize our willpower levels. Still, our willpower is a limited resource. Despite our best efforts, we will all reach points where we are depleted and levels are low. Simply wishing for more is not a helpful strategy, and it is more important that we learn to use our precious willpower resources wisely and efficiently.

One of the biggest mistakes people often make is ignoring willpower constraints and going for broke. This risks us becoming overwhelmed, so we give up, only to end up repeating the same cycle over and over. As a society we are obsessed with willpower and motivation, and we adulate large actions. We are drawn to things requiring a complete re-invention of self or a major

---

* This cycle does not affect everyone equally: there are people whose executive battery has reduced capacity, perhaps due to a neurodevelopmental condition such as Attention Deficit Hyperactivity Disorder (ADHD) or another neurological illness developed later on in life. The executive is able to compensate for deficits in other parts of the brain but, in doing so, this drains executive reserves and affects overall capacity. These individuals will be more susceptible to entering low power mode, sometimes even constantly operating in this state. Some of this impact can be alleviated by implementing techniques in this book that use other brain regions to support the executive.

upheaval. We want to create changes in our lives by sheer brute force but when the power of our will inevitably fails us, it in turn becomes our own personal failure.

The reason you pick up your phone is not because you lack willpower or are unmotivated. Picking up this book and wanting to learn more about the subject indicates the opposite. You have instead developed a set of digital habits that have now become problematic and difficult to control. The same properties of habits that mean we don't waste valuable brain power mean they form our default option at times of mental strain when we are disengaged or fatigued. Rather than using our executive to fight against it, we need to change our default so that it supports our goals. As we discovered in the previous chapter, habits play a pivotal role in pre-selecting our choices so, in essence, establishing habits that align with your goals provides an effortless and non-fatigable method of self-control.

Low power mode is something that you may have recognized in yourself as you have been reading this chapter. However, the signs are not always easy to pick up. The participants doing the complex letter recall task did not. After six hours of complex cognitive tasks and despite reduced activation of their executive brain, they themselves did not report increased levels of exhaustion.[27] This is not surprising. As the executive steps back, we lose the ability to critically analyse our internal state. We lose insight. Adequate rest and restorative breaks are essential for maximizing willpower and avoiding the development of problematic habits – whether you feel the need for it or not.

However, our individual habits cannot be viewed in isolation; they are influenced by a broader perspective that must be discussed. Although humans have a uniquely developed executive compared to other animals, who simply make decisions surrounding breeding and surviving, our brain evolves slowly and the demands of modern life have vastly outpaced our capacity. Unrealistic expectations have become the norm, with long working hours, minimal breaks, and insufficient sleep becoming the standard. We chronically overuse and undercharge. Consequently, many of us find ourselves operating in low power mode much of the time, leading to inefficiency and suboptimal performance. In this cognitively depleted state, we make impulsive decisions and develop contradictory habits that become coded in our autopilot and then require significant willpower to override. To address this issue, we must go beyond individual efforts and challenge societal norms and unrealistic expectations. This is one of the key motivations behind writing this book. While transforming deeply ingrained societal expectations and norms may require a collective effort over time, the upcoming practical section equips you with actionable strategies that allow you to kickstart meaningful changes in your habits and enhance your brain health, starting right now.

Part I Practical

The Foundation Toolkit

We often check our phones on autopilot, having stored a variety of digital habits, some of which may be problematic. To begin the process of changing our habits, we need to interrupt autopilot mode and engage the executive brain instead. However, given the demanding nature of our lives, most of us are operating at the limit of our mental energy, and have a reduced capacity to make changes.

Being in low power mode does not mean that you are powerless. All the techniques in this foundation practical section take our executive function constraints into account. They are designed to help you start the process of change but can also be used in low power mode. I've used them countless times myself, including in the creation of this book.

I placed them here to make the most of your motivation so, rather than waiting until the end of the book, you can start applying them straight away. Some of the strategies we will explore are standalone approaches, applicable in specific circumstances. However, others are foundational, and are labelled as building blocks, as we will continually develop and expand upon them in subsequent sections.

Take Stock of Your Phone Habits

Look at your digital habits and consider which ones you'd like to change.

- How much do you want to change and why?
- At what times would you prefer not to be on your phone?
- What would you like to do instead of scrolling on your device?

★ Remember: do this in a non-judgemental way as there is no single ideal way to use technology. For one person, checking their email first thing in the morning might mean they can mull something over on the way to work, while for another it could be a source of stress.

*

List your habits in a table under the headings: supportive, problematic and neutral.

- Be specific: rather than a vague focus on screen time, detail the occasions when you would like to limit your phone use, for example, when you're trying to study, watching a film or at the park with your children.

- Start small: taking on too many things vastly reduces your chance of success. Small changes when implemented early and consistently can frequently outweigh an attempt at a drastic overhaul.

- As your relationship with your phone changes, it is important to decide how to fill the new gaps in your time, otherwise you might revert back to your phone again.

Building Block 1: The Five-Minute Rule

When faced with the urge to check your device, delay rather than resist. Pause for just five minutes.

- The need to do a 'quick check' of your phone often disguises an urge to escape a present situation. Most of the time that urge is momentary, but nonetheless the autopilot brain triggers a habit sequence.

- There is no mind-trickery involved – after the five minutes have elapsed, you are welcome to pick up your device. And whether you do or don't, you should consider it a victory.

- Often, this small pause is enough to stop you from being sucked unnecessarily into the virtual world and, even if you do end up checking your phone, you have given your mind the opportunity to sit with some discomfort and to think about how to deal with it.

- A five-minute delay prompts your executive brain to step in. This means that when you do reach for your phone, it is a mindful act rather than an automatic response.

- While strict rules, such as completely abstaining from phone use, may seem more effective, they demand considerable willpower and are prone to crumble when you're in low power mode, especially if the habit has not been built.

Tips:

- You don't have to do this all the time. Identify specific situations where you would benefit from having this rule, perhaps when undertaking important work, or when spending time with loved ones.

- If there is a good reason to check your phone, add this to a list of things to do intentionally or look up later.

- Remember, start small. Begin with one specific scenario and, when you're successful, extend it to other areas.

Surf the Urge

It is a common misconception that the urge to do something increases over time. This is not the case. Habitual impulses tend to come, peak and then dissipate like a wave. 'Surfing the urge' is a technique used to help people deal with smoking cravings* but can be applied to many other situations. It consists of mindfully observing the sensations that arise when a craving occurs. Instead of acting on an urge, you surf the wave, noting down how your body and mind feels, until it eventually disperses, similar to a wave hitting the sand.

- To gain insights into why you instinctively reach for your phone, leverage the delay strategy of the Five-minute Rule. During this pause, assess the spectrum of emotions that arise and jot them down on paper.

- Over time, you'll likely notice patterns. Do you feel restless, uncomfortable, sad, or are you attempting to procrastinate a challenging task?

- Identifying these feelings shifts them from our subconscious to our conscious brain, allowing us to rationally evaluate them. Remember, as humans, we possess the unique ability of metacognition - we don't just think and feel, we can analyse our own thoughts and feelings. Being aware of why you feel compelled to reach for your phone is the first step towards change and will equip you to apply the strategies in this book more effectively.

---

* Although nicotine is an addictive substance, the act of smoking consists of many habitual behaviours that closely echo phone use. Nicotine substitutes such as patches or chewing gum can address the chemical dependency, yet people often still experience a strong urge to reach for a cigarette. This compulsion is akin to our instinctual tendency to pick up our phones, and underscores how both these actions can become deeply ingrained. This is precisely why smokers often express the need to 'kick the habit'.

Building Block 2: Plan B

Have a Plan B for low power mode to avoid defaulting to Plan 0.

- We often set Plan As full of ambitious goals, only to berate ourselves when our motivation inevitably ebbs at some point, and we do not achieve them. We do not exist in a perpetual high-motivation state, so when you slip into low power mode, deploy what I like to call 'Plan B' – this is a technique I personally created to keep myself on track and which I use a lot.

- A good Plan B should be an activity that (a) aligns with your overarching goals, (b) leaves you with a sense of achievement but (c) is less challenging for your executive than your Plan A.

- Plan Bs are underused because we mistakenly equate them with failure. Having a Plan B demonstrates an awareness of your brain's fatigue levels. Lacking a Plan B breeds an all-or-nothing mentality that could lead to automatically defaulting to Plan 0 – such as binge-watching content or aimlessly scrolling through your phone.

Some examples:

PLAN A	PLAN B	PLAN 0
Working on a complex project	Doing admin	Getting distracted on the Internet
Learning a new topic	Revising an old topic	Refreshing Twitter
Studying	Organizing notes	Scrolling through Instagram
High intensity workout	Going for a walk	Watching YouTube
Writing a paper	Passive reading	Checking the news

★ Remember: unlike our phones, there are no alerts to signify when our brains are in low power mode. If you find that your Plan B isn't quite cutting it, chances are your executive brain is drained. In this state, taking a genuine break to recharge might be the most productive decision. Even a few focused, productive hours often yield better results than spending an entire day in a perpetually distracted state.

The Signs of Low Power Mode

SIGN	WHY
Feeling unmotivated	The executive brain is fatigued
Less able to focus	
Having difficulty in making easy decisions	
More likely to act on old habits	The executive has delegated to the autopilot brain
More likely to reach for short-term rewards	
More likely to procrastinate	
More irritable	The fatigued executive is having trouble regulating the emotional brain
More prone to engaging in negative self-talk	
More likely to lose temper	

MAINTAINING CONSISTENCY

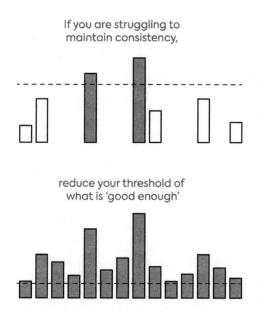

If you are struggling to
maintain consistency,

reduce your threshold of
what is 'good enough'

How Plan B Builds Habits: If you are constantly aiming for an ambitious Plan A but struggling to maintain consistency, reduce your threshold of what you deem 'good enough'. Having a Plan B increases consistency, which will eventually programme new habits.

Building Block 3: Insert Hurdles

Use the power of hurdles to disrupt autopilot mode so that you check your phone intentionally, rather than habitually.

When the autopilot brain encounters a hurdle, the executive is activated. This means you have to stop and think – do I really want to check my phone?

Handset tactics:

- Place your phone beyond easy reach so that you have to make a conscious effort to stand up or shift position.

- Switch your phone off. A sign of autopilot mode is that you reach for it without thinking only to be surprised when it doesn't respond as expected. Having to switch your device on and off each time you use it creates a tangible delay which activates the forward-thinking executive.

- Stow your phone at the bottom of your bag rather than an easily accessible pocket. Pair this with switching off your phone for an effective double hurdle. Doing this when you go out provides the security of having a phone with you, which many of us feel we need, but without it being readily available at a moment's notice, leaving you free to immerse yourself in your surroundings.

App approaches:

- Log out of selected apps after use. Do not set touch ID, face recognition or another method of quickly inputting your password. Remembering your password requires your executive to be activated so you are less likely to act on autopilot.

- Two-factor authentication is not only great for security but also slows down the login process, giving your brain a chance to reconsider and make a different choice.

- Relocate apps associated with problematic habits away from your home screen and into a folder. If you find yourself automatically navigating to their previous location, you've spotted an autopilot action and it's an opportunity to use that pause to apply the Five-minute Rule (Building Block 1).

- Delete apps and re-install them each time you want to use them. This is a more powerful hurdle for those who feel they need it due to having developed problematic habits.

★ If you check your phone despite these hurdles, it does not mean they didn't work. Rather, shift your focus from your next phone check to your next thousand checks. If the average person checks their phone close to one hundred times per day, that number could be reached in a mere ten days. Although small, these hurdles have the power to shift the balance of our actions - perhaps to 500 checks. You'll be amazed by how much a small dose of friction can help to cut down on those unnecessary phone checks while still letting you use your device when needed. And the best part? It allows you to strike a balance without having to resort to an all-or-nothing approach.

*

★ A note of caution when it comes to these hurdles. The autopilot brain is efficient and is able to form new pre-programmed sequences that incorporate your new hurdles, so you can end up back in autopilot mode again. For instance, typing your password multiple times a day can quickly become part of a new habit. Capitalize on the initial disruption these hurdles provide and combine them with the other tools offered in this book to maintain their effectiveness.

Building Block 4: Pre-commitment

Capitalize on willpower highs to plan for willpower lows.

Pre-commitment means strategizing ahead of time and using periods of high mental energy to craft a plan for when your will-power reserves dip. This forward-thinking approach grants you control, even as your willpower wanes.

Tactic 1 – Limit your options in advance:

- Move your phone away from your bedside to pre-commit to curbing your early morning scrolling.

- Leave your primary phone at home while carrying a basic one with only essential contacts when you study at the library.

Tactic 2 – Set implementation intentions:

- Work on the simple principle, 'If X happens, then I will do Y.' This is a psychological strategy proven to bolster goal achievement. Having thought of a strategy in advance, when your mental energy is high, will take some of the strain from your executive brain when in low power mode. This can prime you for constructive reactions when challenges arise.

- You could pre-commit which tools from this book you will apply in which scenarios. Keep a list like the one below which will get longer as you go through the book:

IN THIS SITUATION	I WILL USE THIS METHOD
Urge to mindlessly browse the web when studying	Five-minute Rule
Scrolling through work emails while in bed	Insert Hurdles (log out of work email account or delete completely from phone)
Checking phone during family time	Insert Hurdles (phone switched off and out of reach)
Waking up and immediately reaching for the phone	Implement the Five-minute Rule to get up, open the curtains and start getting ready
Not in the mood for exercise and tempted to binge-watch content	Plan B (go for a walk or do some stretches while listening to a podcast)
Constant refreshing of a social media app	Insert Hurdles (log out after each session, move app into a folder, delete app)
Unwanted online shopping sprees	Insert Hurdles (do not save passwords, do not store card details)

Setting New Patterns of Phone Use

Start delegating healthier digital habits to your autopilot brain.

- Use pre-commitment to establish what a balanced pattern of phone use will look like for you. Rather than pre-committing not to check your phone, pre-commit to when you will positively and intentionally engage with it.

- Set aside time for intentional technology use such as catching up on news, planning what content you will stream and catching up on social media. Aim to create a balance with other activities such as exercising, work, admin and social interactions. It is important not to be too rigid; leave some time free where anything goes.

- Building this structure means eventually your brain will no longer navigate towards those apps outside of those situations – much like we only have set times of the day to brush our teeth.

- A more flexible approach would be to pre-commit a maximum limit of how often you check an app within a day but without exact times. This avoids short, frequent, distracting checks and encourages intentionality, anticipation and enjoyment.

★ Remember: Do not be overly restrictive as this is likely to end in failure and disappointment. It is far better to succeed in checking an app once per hour than to fail at checking it twice per day. Remember that small successes can be easily built upon. The exact number of times you pick up your phone is less important than choosing intervals which are short enough to require little or no willpower, meaning you can maintain this even in low power mode.

Strategy

Capitalize on Change

When life provides you with a blank slate, use it as an opportunity to alter your phone habits.

- If you're due to undergo a major life change that will lead to a disruption in your routine, then double down and implement a new technological routine too.

- Major life changes are relatively infrequent, but you can try and get the same context change in smaller ways: going on holiday, the start of a new academic year or starting a new job.

- Designate a new location such as a coffee shop or a library - a space that your autopilot has not linked to any existing digital habits around - and use it as your 'ground zero' to create new digital habits.

- Capitalize upgrading to a newer model of phone - as your brain adapts your habits to a slightly different layout or configuration of buttons, you can use that switch to optimize your phone habits.

- Milestones such as the new year, the beginning of each month or even the start of the week offer bursts of motivation. They signify the passing of time and prompt us to focus on the broader picture, rather than being bogged down by daily minutiae. Leverage these junctures as opportunities to employ the strategies outlined in this book.

- Any insights or a-ha moments you encounter in this book can also boost motivation and create a sense of a new beginning. Use these as launchpads for your journey towards healthier digital habits.

Willpower Rules

- Willpower depends on the executive brain. It will fluctuate depending on how cognitively fatigued you are.

- Sleep is the most powerful universal method of recharging willpower.

- Having adequate nutrition and taking energizing breaks slows down the rate of depletion.

- When the executive brain gets fatigued, your brain enters low power mode.

- In low power mode, you may form contradictory digital habits which then deplete your willpower further.

*

Congratulations on completing Part I. You now have a better understanding of how your brain works and why you reach for your phone. Now it's time to move on to the next step.

We are more likely to form contradictory habits surrounding our phones than other objects because of certain qualities our devices possess. In Part II – The Habit Puzzle – we will focus on reprogramming our brain's autopilot system and you will be equipped with the knowledge of how to form new supportive habits and rewire contradictory ones. Arming your autopilot with habits that work towards, rather than impede, your goals means you support your executive and maximize your potential. It's like having extra brain power.

PART II

The Habit Puzzle

4

Deconstructing Habits

If making a change in our behaviour merely depended on our wish to do so, then this book would not be necessary. The extensive negative media coverage surrounding technology would have already led to widespread changes in people's digital habits – except that it hasn't. Knowledge of potential adverse outcomes is not enough to change habits, and it is common practice in medicine to see people with habits so powerful that they struggle to change them even when faced with life-threatening health information. This is because frightening someone into changing is not only ineffective but can be counterproductive. Our executive brain might contemplate future consequences, but our habits reside in the autopilot realm. Although the executive, being the boss, can override the autopilot, this action is akin to swimming upstream – exhausting and unsustainable in the long run. Further, the added emotional strain that such fear-based tactics place on our executive only serves to undermine the process.

Many people try to control their habits in a way that's similar to how we control our breathing, which is another process that occurs largely outside our conscious control. Breathing is managed by our brainstem, a critical part of the brain that connects to the spinal cord. While we can consciously manipulate our breathing rate and pattern using our executive brain, this can only be sustained for a limited amount of time before something else requires our attention. Trying to control our breathing continuously would be too energy draining, as is attempting to exert rigid control over our digital habits. Our executive can only enforce such strict rules for so long before we find ourselves in low power mode and unable to spare the excess mental energy.

The digital detox fallacy

Changing phone habits in many people's minds is synonymous with going on a digital detox, something that has received a considerable amount of press attention. How a digital detox is applied varies but it essentially means not checking your phone or a particular app – usually social media – for a predetermined amount of time ranging from a single day to a month or even longer. Many doctors, including myself, are against the term 'detox' because it is often associated with unscientific and alarmist claims about removing toxins from the body, a concept that doesn't apply to technology habits. Our phones are not toxic, and we are not detoxifying. Setting aside any reservations about the wording and potential false claims, I am fully supportive of people spending time away from their phone should they choose to, but there a few things to note.

It's important to recognize that not everyone has the privilege of being able to disconnect from their digital devices. Many people rely on their phones for work, to remain in contact with their loved ones and, especially if they are parents or carers, they need to be easily contactable. The good news is that you do not need a digital detox to change your phone habits and, in fact, such detoxes may actually be ineffective. In addition to the impracticalities of going on a digital detox, treating our phones like a drug and relying solely on abstinence as the solution has some major pitfalls. While abstaining from phone use for a day might make you more aware of your habits and inspire you to reconsider your relationship with technology, on its own it is unlikely to lead to any meaningful long-term change. For most of us, our phone habits have become so deeply ingrained in our autopilot that they will likely need several weeks or even months to be rewired.

The same goes for longer digital detoxes. Even when someone disconnects for several weeks, it is typical to return to old patterns of phone checking when the offline period is over. This is because the all-or-nothing nature of a digital detox provides a sharp cut-off point. You go all out to remove technology but when the detox is finished you revert back to nothing. Think about it this way. Going on a fad diet, where you fast, won't teach you the supportive habits you need to eat healthily, buy and cook nutritious food, or make balanced choices about your diet. Similarly, completely switching off from technology won't teach you how to use your phone in moderation. This means that, at the end of the abstinence period, you haven't created any supportive phone habits in the autopilot brain, which reverts to its previous programming – your old habits. Moreover, most digital detoxes are done once, and infrequently repeated events do not become habits. A digital detox is like renewing our car insurance, an event to be done yearly, whereas to form supportive habits we need constant daily actions, the equivalent of brushing our teeth. You should therefore be aiming to incorporate aspects of the tools provided in this book into your daily or, at the very least, weekly routine.

The four elements of the Habit Puzzle

The reason we try these dramatic disconnections from technology is because, when making changes, most of us feel the need to start big. We need to feel like we are working hard, and that we are achieving something. While small changes might seem unambitious compared to the perceived discipline required for a complete overhaul, it's crucial to understand that ambition should not be mistaken for impatience, which often underlies such lofty symbolic gestures.

To make sustainable long-term changes, it is OK both to start small and to focus on the changes that feel the easiest. Whenever I explain this to my patients, I use the analogy of completing a large puzzle. When doing complex puzzles with many pieces, most people start with the edges or choose a part of the picture that stands out. Both of these techniques make the puzzle easier to complete. And here is the key: as you build this puzzle, the pieces that once seemed difficult to place start to fall into place much more easily than they would have at the beginning. When the puzzle is completed, the end result is the same no matter what route you took to get there. Starting small does not signify a lack of ambition. Embarking on change is hard and choosing what feels the most achievable route is smart, strategic thinking.

If you can relate to the above and have found yourself repeatedly fighting against problematic digital habits stored in your autopilot, it's time to adopt a new strategy. Our autopilot brain is there to support the executive, and so we need to reprogram our phone habits in a way that aligns with our goals. This is not a task that can be completed all at once; we need to approach it piece by piece, targeting particular problematic digital habits and slowly replacing them with supportive habits, which can be digital or non-digital. To do this, we must first understand how habits work.

There has been substantial progress in our understanding of habit formation over the past thirty years with thousands of scientific papers published each year. Having spent a considerable amount of time delving into this research, I could see that insights cluster around four key themes. They are (a) how our habits are initiated, (b) the types of actions that are habitual, (c) the chemical processes in our brain that drive change and (d) the length of time it takes to form a habit. I've used those key themes as inspiration and, applying my most-used analogy, brought them together in the form of a puzzle with four pieces each reflecting the above themes: Reminder, Really Small Action, Reward and Repetition.

Here is how each individual Habit Puzzle piece works:

1. **Reminder** – This is a trigger that prompts the autopilot brain to perform a habit. It could be something like a certain time, a specific location, or an emotional state.

2. **Really Small Action** – This is what we consider the habit itself. The autopilot is only able to code really small actions as habits because more complex actions still need at least some input from the executive.

3. **Reward** – This is the positive outcome that we receive as a result of the Really Small Action. Rewards release chemicals that help bind the reminder to the Really Small Action.

4. **Repetition** – Consistently repeating the habit over time means that our neural connections get stronger, making the habit easier and more automatic.

The Habit Puzzle: Each one of our habits is a Really Small Action triggered by a Reminder. Reward and Repetition are an integral part of the saving process. Together, these make up the four key components needed to encode a habit.

Each of these elements is needed to programme a habit sequence into our autopilot brain. We usually perceive the first two elements – Reminder and Really Small Action – as the habit itself. Upon recognizing a reminder, our autopilot brain will execute a very small action. The subsequent two elements – Reward and Repetition – are vital components of the saving process. The reward kickstarts the saving process while repetition completes it, thus completing the 'Habit Puzzle'. This leads to the automatic response of reaching for our phone in various scenarios.

Think of each habit as a programming script or a piece of code in your brain that the autopilot uses to automate your behaviour in situations it deems appropriate. We all have a multitude of these completed puzzles in our autopilot, and it's these unique habits that dictate when, where, and why we reach for our phone.

When it comes to the brain, nothing proceeds in an orderly fashion. Everything is interconnected. Elements come together in different orders or simultaneously, and the Habit Puzzle illustrates this. Reminders may be the obvious first step if looking at it like a list, but the brain doesn't work in lists. For example, you may start with doing a really small action and your brain will subsequently link it to the reminders in your environment, building up initiating triggers to make it easier to do the same action next time. An unexpected reward can also jumpstart this habit-forming process. Your brain looks back at what caused the reward, focusing more on what happened before, and saves these as reminders and small actions in the hope of getting the same reward in the future.

Additionally, rather than a physical puzzle, the components of this metaphorical jigsaw are chemical representations between nerve cells. As a result, they are more fluid than standard puzzle pieces and vary in strength. Increasing the strength of the components speeds up habit formation. More reminders to activate the habit means the habit will form faster. The smaller the action, the more easily it can be coded into the autopilot brain. More rewarding habits are saved more quickly, and more repetition strengthens the neural connections in our brain, cementing the habit in place. Weakness in one of the components can be made up by strength in another. For instance, a less rewarding action can still become a habit, but it requires more repetition.

Phone habits are unique in how efficiently they fulfil every single component of the Habit Puzzle. There are reminders everywhere, both external ones in our environment and internal ones in our mental state, that prompt us to initiate a phone-checking action. Our phones have been designed to be frictionless, making it easier for our autopilot mode to execute really small actions. The breadth of the digital world means that there is abundant reward availability. Finally, the average person will reach for their phone several times per hour, a figure that vastly outnumbers any other action in our daily lives.

The interconnectedness of these elements means that undoing problematic habits requires intervention at each step of the process. Therefore, in the next few chapters I will delve deeper into each element of the Habit Puzzle in turn, to give you a better understanding of what goes on inside your head when you engage in a habit. Understanding why we do the things we do is in itself enough to start sowing the seeds of change but, to capitalize on this even further, Part II will culminate in a practical section with techniques designed to target each piece of the Habit Puzzle. This will help you change any problematic habits into supportive ones and essentially reprogram your autopilot mode. Start with the pieces that feel the easiest and the rest will fall into place.

5

Reminders

To everyone else, it was an outdated song playing in the background of a little sailboat on a Greek island, not to be given a second thought. To me, the song represented a reminder powerful enough that my mind was no longer there. I was transported to a different location, to a dark room in the laboratory where I'd spent countless hours looking down a microscope while counting neurons, with the same song playing in the background. The feeling was so strong that I could practically feel the cold metal of the microscope. The bumpy texture of the dials was just beneath my hands. I sensed the delicate movements needed to adjust the slide; a double-click to take a picture. The muscles in my hand prepared to move, the rest of my body remaining carefully poised not to disturb the balance. But the only connection between those two places was in my mind.

We've all had experiences like this. The reason this happens is because our brain is an association machine. It is constantly making connections, and an unexpectedly familiar sight, sound or smell is enough to trigger vivid imagery. But in addition to mental imagery, our surroundings can be linked to physical actions too – this is what habits are. A habit is an automatic physical response to a situational cue, termed stimulus-response by scientists. The autopilot brain depends on these cues, using them as reminders to initiate habits. This is the first piece of the Habit Puzzle.

External reminders

Unbeknown to us, our autopilot brain is constantly scanning our environment looking for external reminders to determine which of our stored habits will leap into action. In the same way music is capable of triggering vivid memories, location will prompt certain habits and is one of the most powerful external reminders that our brain uses. For instance, if a runner is reminded of the location they usually run in, it can automatically trigger thoughts of running and may create a desire to go for a jog.[29] Walking into our bathroom can provide a reminder to our autopilot brain to brush our teeth.

How our autopilot brain uses location-based reminders to execute habits was examined by a group of scientists who closely observed a collection of students eating. The students, under the impression that the study was about how their personality affects their taste in movies, did not realize that the small red and white stripy bag on their lap was the actual experiment. Filled with popcorn, the bags were discreetly numbered and carefully weighed so the researchers could measure how much popcorn the students consumed throughout the experiment. What they found confirmed their suspicions. How much the students ate depended on the location they were placed in. Students in the familiar setting of a cinema ate much more popcorn when compared to another group of equally hungry students who did the same experiment sitting in a conference room.[30]

Our decision-making power is too valuable to be spent considering small, repetitive actions like reaching inside a bag of popcorn, and checking our phone is similar. Having a popcorn bag on our lap, or our phone nearby, can trigger habits but our surrounding environment also plays a big role. Being in a particular setting can remind our autopilot to reach for our phone while a different environment will not – the difference between the cinema and the conference room. This process happens subconsciously so, in the right setting, the popcorn bag in your lap will get progressively emptier in the same way that, without realizing, you are more likely to reach for your phone in specific situations.

Apart from location, time is also a powerful external reminder that can trigger our stored habits. Most people brush their teeth each morning and evening with no special effort to remember. Our autopilot combines location and time to activate the correct habit, so we don't get the urge to brush our teeth outside our bathroom or when walking into the bathroom in the middle of the day. The combination of time and location also governs many of our eating habits. Studies show that we make over 200 food choices per day, most of which are habitual,[31] meaning that we tend to eat the same or similar foods at the same time and place. Someone may have no desire to eat a chocolate bar at breakfast, but if their autopilot has stored a habit of eating something sweet after dinner, that chocolate bar in their cupboard will suddenly become irresistible around that time.

This combination of location and time plays a crucial role in our digital habits, including any problematic ones. Prior to the availability of smartphones, the habit-forming nature of the Internet and computers was kept at bay by a key feature: it was location-restricted. Just like you generally only order (and possibly end up overeating) popcorn in the cinema, you could only check your computer while at your desk. Computers eventually became smaller and more portable in size in the form of laptops, but they were not mobile enough for many situations. You wouldn't casually pull out your laptop at the dinner table and begin checking it as soon as the conversation became a bit stilted. The limited number of location- and time-based reminders naturally restricted our digital habits, allowing for long natural breaks during which people engaged in other tech-free habitual activities such as commuting, eating and exercising. The advent of the smartphone changed that.

Having pocket-sized devices that could connect to the Internet with the processing power of the computers I grew up using meant we started carrying them everywhere. This unlimited access provided opportunities to form habits beyond the confined locations

74

associated with our desktops and laptops. Our phones have likely touched every surface in our house and very few other objects in our life have been granted the same privileges. Every time we use our phone at a particular place and time, we are training our autopilot brain that, when similar reminders are encountered, a phone-checking sequence should be initiated. Just like if you were to carry a bag of popcorn constantly, you would start to form popcorn-eating habits beyond the cinema.

Unrestricted time and multiple locations provide a substantial number of reminders associated with phone checking, but so can the phone itself. Just like seeing a plate of cookies will make you want to grab one, seeing our device can lead to automatically picking it up, as can seeing someone else check their phone. Additionally, unlike other habit-forming objects in our environment, our technology is not passive, and smartphones produce their own reminders. Notifications or alerts, familiar pings, and buzzing sounds all automatically capture our attention, even when they are not played from our phone. The instinctive response to hearing another person's similar ringtone reveals the links that have formed in our autopilot brain which usually proceed without our conscious awareness. Scientifically, this is termed 'cue sensitization', meaning that our brain has become sensitized to that particular reminder. The apps on our phone also act as digital reminders. Unlocking our home screen means that we are greeted with a range of familiar icons which prompt us to do specific actions more reliably than the general Internet browser, a hit-and-miss method that relies on a person remembering what they wanted to look up. It's this increased number of reminders that can serve as the first piece of the Habit Puzzle that ultimately accelerates the formation of new phone habits compared to other habits in our daily lives.

*

The building of habits that were unrestricted by location was one of the keys behind the success of the social media platform Instagram. Exclusively built as a mobile app which launched in

2010, it was inaccessible from any desktop until October 2012. The app was prioritized, and the desktop version was clunky. Over ten years later the Instagram website still suffers from limited functionality compared to the app. To a casual observer, this may appear to be a mistake, an oversight of sorts – except, it is not.

When Instagram was launched, its direct competitors and more established social media companies Facebook and Twitter had spent years honing the social media habits of a generation. People had already established habits of frequently checking their social media feeds, sometimes in a way that they found disruptive. The aspiring company that was Instagram was able to capitalize on their hard work and build habits with one key difference: location. While both Facebook and Twitter had created earlier apps, you did not have to download them onto your phone to be able to access those social media sites and could continue checking them in a location-dependent manner on a desktop or laptop. In contrast, signing up to Instagram meant installing an app which you could check regardless of time and location. The increased number of reminders built stronger habits in the autopilot brain and, as a result, by 2020, half of Instagram's one billion monthly-active users checked Instagram daily,[32] many even several times per day, having truly integrated this into their everyday routine.

If you've ever felt like you lack space from your phone, this is why. As we accumulate more situations where we check our phones, we create a complex web of reminders persistently triggering the autopilot brain to activate our digital habits. Done often enough, this begins to infiltrate every aspect of our routine, where we constantly and automatically reach for our phone and feel that we can't truly disconnect. One of the keys to creating balanced habits, which will be discussed in the Part II practical section, is to unwind some of this intrusion, consciously reducing the number of situations in which we check our phones in order to diminish and weaken the network of reminders that are triggering our autopilot brain.

Internal reminders

A common piece of advice that is offered to people who want to rethink their phone habits is to turn off notifications. We think that we are disrupted by constant alerts caused by the device itself but, most of the time, this is not what happens. In Chapter 2, we discussed a study conducted by the London School of Economics. For this research, participants were asked to wear glasses equipped with a mounted camera, providing researchers with a direct viewpoint into how they engage with their smartphones. After thorough analysis of the captured footage, a surprising revelation emerged: in 89 per cent of instances where participants checked their phones, there was no notification prompting them to do so.[19] They instead reached for the handset seemingly out of nowhere, and it is those unprompted checks that account for the majority of our phone checks. The way we use our phone has become so habitual that we reach for it every few minutes whether there has been a notification or not. Our habits proceed in a stealth-like fashion and are much less noticeable than a buzz or alert, making us think that other factors are to blame for disruptions when we need to look closer to home.

Not only has our autopilot been programmed to respond to external cues in our environment, but, in many instances, a strong urge to check our devices comes from within. It is not unusual for a behaviour that was originally triggered by external reminders to develop internal reminders over time – and we cannot control which associations our brain forms. An external physical trigger, such as a hangnail, may lead to someone biting their nails, but once the behaviour is repeated in several situations, the brain starts to link it to an emotional state. It is not long before anxiety, impatience or frustration become the internal reminders for someone to absent-mindedly perform the same behaviour in the absence of the physical trigger.

Using our phones in an unrestricted fashion means that, in addition to external reminders, our brain machinery starts to link our habits to our internal state. When we encounter the same internal state in future situations, we reach for our phone – this physical action becoming a universal outlet for our emotions. Our phones can provide temporary relief from a lot of uncomfortable emotions, ranging from outrage to sadness and boredom; a big digital distraction plaster that we put on our wounds to grant us temporary reprieve. Sometimes this modern-day coping strategy can be useful. The passing of time helps emotions fade, we may gain perspective, we may reach out to a friend for emotional support. Incorrectly applied, the strategy may prevent us from taking the active steps needed to improve our situation, something which will be discussed further in Chapter 11, Mental Health.

The common recommendation to limit notifications in order to reduce the habit-forming nature of smartphones is something I support. Sometimes notifications can form the initial external trigger – the equivalent of the hangnail – for our habits, plus turning off non-essential alerts is easily done and requires little willpower. However, it is important to be mindful that this is a rather simplistic piece of advice for what has now become a complex problem. Some studies show that even drastically reducing notifications to zero does not seem to have any impact on the overall pattern of smartphone use.[33] In my experience many people have a habit of ignoring notifications – especially when they are numerous, leading to notification fatigue – but continue to reach for their phone completely unprompted. When someone's phone use has become problematic enough to encroach on their daily life, it is likely that they have formed strong habits linked to a multitude of external and, most importantly, internal reminders so further strategies beyond turning off notifications will be needed.

It's crucial to recognize that our phones have, at times, evolved into a coping mechanism. However, it's equally significant to approach this without judgement. This is where the Five-minute Rule, as well as its advanced version Surf the Urge, come into play. Introduced in Building Block 1, these techniques offer your brain the opportunity to digest uncomfortable feelings in a manageable, controlled manner. The Five-minute Rule allows you to delay the automatic response of checking your phone, fostering the development of alternative coping and self-soothing strategies that won't feel too overwhelming. Just five minutes can create a gap between the reminder and a small action, helping to dismantle the first two pieces of the habit puzzle, and consequently problematic habits. As you progress and become more comfortable with this pause, you can begin to Surf the Urge – acknowledging and observing your feelings and the urge without immediately acting on them. These practices are crucial as they empower you to navigate discomfort mindfully, eventually allowing more adaptive responses to surface.

Stopping reminders

We often use our phones in a way that helps our daily tasks; we may need to check the time, quickly look up information or send someone a message. However, the association machine that is our brain often thinks in a tangential manner. Irrelevant thoughts frequently pop into our heads, not all of them important: a task that we've been meaning to do, a message that we've not yet replied to, the name of a movie that we can't quite remember. Rather than letting these irrelevant thoughts pass by, or noting them down if they are actually of value, we can now couple them to a physical action. We can pick up our phone. This is more likely to happen if we are in low power mode, bored or distracted. We begin to follow this train of thought, entering the virtual world like Alice in Wonderland, except that, when we end up in a scroll hole, it is much more difficult to exit.

As well as initiating reminders, our daily life features a multitude of stopping reminders. These provide a terminating sequence that enables our autopilot brain to know when to end our habitual action. This pause also gives us time to think about what to do next. Pauses activate the executive brain, which decides our subsequent action and, without them, our autopilot brain will continue along the same course. Just like internal reminders, stopping reminders are also subdivided into internal and external. For instance, we may either finish eating when we feel full – an internal reminder – or when our plate is empty – an external reminder. External reminders, like the empty plate, are more reliable and can sometimes overpower our internal reminders. In a study where researchers secretly switched people's plates to a 50 per cent bigger size, they still ate the same percentage of food they normally would due to an overreliance on the external cue of how much appeared to be left on their plate.[34] Most people reading this book will have, at some point, eaten past the point of fullness – finishing a big, rather than a small, pack of crisps, or polishing off a tub of ice cream. We ignore our internal signals of satiety and fullness, relying only on external reminders to know when to stop.

An abundance of initiating reminders leads us to frequently check our devices, but we will put them down quickly when nothing new is encountered: a lack of new messages, news stories or emails provides a powerful stopping reminder. However, the success of tech companies depends on capturing our attention and as a result, many apps have been designed to lack as many pauses or stopping reminders as possible. For example, when social media sites still relied on a chronological social media feed, quick frequent checks meant that the typical user would quickly scroll down to the point they had encountered previously and, seeing the same content, stop reading. That sense of completion was a powerful stopping reminder. The introduction of endless algorithmic social media feeds, which constantly deliver new content, is the equivalent of a virtual bottomless ice-cream tub. The autoplay feature, which automatically starts the next episode of a TV show, eliminates the need for our executive to decide whether to continue watching. Instead, we may effortlessly follow the path of least resistance, resulting in longer viewing times.

Unlike the stretch receptors in our stomach which signal to our brain that we have eaten past the point of fullness, binge-watching and mindless scrolling do not have a reliable neural system that leads to satiety. The disproportionate effect of having many initiating reminders and relatively few stopping reminders makes it easy for us to pick up our phone but considerably more difficult to put it down. This can happen to the extent that we feel that we are losing control; rather than actively stopping use of an app, we get distracted by another online activity that is also vying for our attention, bouncing from app to app or from link to link, entering ever deeper into a virtual scroll hole, all the while activating our neural pathways and forging new habits that mean we are more likely to repeat this behaviour again. Over and over and over again.

Social media companies, in particular, have come under fire for their never-ending algorithmic feeds but they have also been given the responsibility to fix this. Pop-ups which inform you how much time you've spent on an app are their solution. These can be effective if you always strictly adhere to them, but it only takes a few dismissals for a new superimposed habit to form in our autopilot system. We continue our scrolling habit and automatically dismiss any pop-ups that appear, making them a minor inconvenience rather than a helpful tool.

To regain control, it's crucial to reintroduce stopping reminders whenever possible. A straightforward practical step is to disable autoplay on all your content streaming platforms. This gives your executive brain the opportunity to decide whether to continue watching, instead of automatically progressing to the next episode or video. If you're prone to binge-watching, this small shift can make a significant difference. To further strengthen this approach, consider implementing the Five-minute Rule (Building Block 1) between episodes. Use this break to stretch, jot down thoughts in a journal, or even do some minor household chores. What you do isn't as important as taking the break itself, and you might find that once the five minutes are up, you're content to wait a bit longer. If it's challenging to resist the pull of a cliff-hanger, try setting a timer for the next episode and watching only the first few minutes – cliff-hangers are typically resolved within the initial five to ten minutes of the following episode. By doing so, you can refine your internal stopping cues, effectively teaching your brain to recognize story closure and understand it's time to step away from the screen.

If you choose external timers or pop-ups to act as stopping reminders, it is important to never dismiss them so as not to create a superimposed counterproductive habit. If this type of habit has already formed, you may find it beneficial to disable these reminders for now, turning to other techniques outlined in this book. You can reinstate them later once your brain has had some space from this habit. When these notifications appear, I suggest implementing the Five-minute Rule – instead of abruptly ending your activity indefinitely, tell yourself that you will simply take a brief five-minute break, with the option to return to your phone afterwards. This small, manageable step doesn't demand a lot of willpower and can be implemented even when you're in low power mode. While it might seem insignificant at first, by consistently applying the Five-minute Rule, you nurture a positive habit that will flourish over time, offering an alternative to the detrimental pattern of mindlessly dismissing incessant notifications.

Changing the context

Each of our habits is associated with multiple external and internal reminders, some of which are obvious while others are subtle, and the autopilot brain uses a combination of these reminders to activate each habit. This combination is commonly referred to by scientists as 'the context'. The context plays a crucial role in executing our habits, but not in a binary, all-or-nothing manner. Rather, specific contexts increase or decrease the probability of engaging in a particular digital habit and even slight differences in context can cause a shift in our behaviour.

Consider this example: imagine being in your living room in the evening, with the time and location serving as external reminders. If you find yourself feeling upset, an internal reminder, you may scroll on social media as a way to distract yourself. However, if your mood differs and you feel more inclined towards playfulness, you might opt to play a game instead. Even if you end up accessing the same app, even minute changes in context can result in different actions – taking social media as an example, this may affect the type of content that captivates you, the way you respond to it, the accounts you seek out, and even the way you message your friends. It's important to note that while our executive brain holds the potential to override these habits, many of these actions operate on such an automatic and subconscious level that they often bypass our conscious awareness. Furthermore, the amount of influence the executive exerts on the autopilot depends on its state – whether it is fully charged and capable or running in a low power mode.

When contexts change, our habits change. We have a different routine on days where we are working versus days where we are not. This is because our autopilot brain creates a new set of habits to suit that particular context. Encountering a completely new context, such as going on holiday, has the potential to weaken a large number of the initiating sequences which trigger our established habits, due to changes in our external environment as well as our internal state. It is not unusual for our routine to be disrupted when presented with an unexpected situation or when in an unfamiliar environment; for example, when we're travelling, we might 'forget' to do things that we've been doing for years. However, to be completely accurate in the true neuroscience way, we never did forget – there was just no reminder for us to execute that particular habit.

Certain life events – like holidays, but also moving house, getting married, moving in with a partner or starting a new job, school, or university – change our context significantly, leading to a major disruption of our routines and stored habits. As a result, they create a blank slate of sorts in the autopilot brain while it figures out a new set of habits for our new situation. Rather than moving in a slow, step-by-step manner, these events provide an opportunity to overhaul a large number of habits and build new, more beneficial ones for our goals. If you are due to undergo a large life change, you should take advantage as it will be much easier to implement the strategies that this book provides.

It is important to note that not all context changes are under our control or an opportunity to build better habits. One of the greatest changes of context is becoming a parent and most parents will attest that their new responsibilities involve a seismic shift in habits. Every single aspect of an established routine is completely overhauled and now revolves around looking after a little person with unpredictable needs. As a result, a lot of new parents might find themselves not getting dressed for the day or brushing their teeth until late afternoon. Looking after a little human and the worry that comes with it also puts strain on the executive. At the same time, not being able to replenish their executive brain due to a lack of sleep means that parents find themselves in low power mode a lot of the time. There may be situations when parents have little else to do, such as when holding a sleeping baby, so they reach for the nearest object, their phone, in an attempt either not to fall asleep themselves or to look up advice. Unfortunately, as a result, some parents find that the number of times they check their phone increases after having children and, once coded into the autopilot brain, it has the potential to become a long-lasting habit.

Pandemic acceleration

Our habits change during different phases of our lives, but there is one situation we all experienced collectively, which led to a rather momentous – but unintended – shift in our technology habits: the global COVID-19 pandemic. The amount of time we spend on our phone had been steadily increasing over the past two decades but when, in early 2020, governments across the world issued 'stay at home' orders to reduce the spread of COVID-19, they were saving lives but they also triggered an acceleration in digital habits.

The pandemic instigated a significant shift in context as lock-downs drastically altered the ritual of people's daily lives. Huge numbers of people were required to work from home, disrupting their usual, habit-based morning routine. For some, their autopilot brain was so confused by the context change that, when they sat down at their new home desk, they realized that they had forgotten to brush their teeth. For others, opting not to get dressed and stay in their pyjamas, clothing typically associated with tiredness, led to feeling sluggish all day. Lockdowns reduced the number of reminders for many non-technological habits, as people round the world were prevented from taking part in outdoor activities. At the same time, cues for technological habits increased given that our screens became our main means of communication, and also one of the key methods of keeping up with the news in such an unsettling and frightening time.

Any event that leads to a high degree of uncertainty can have a powerful impact on our brain, such as considering threatening news stories or feeling anxious about medical test results. Given the future-oriented nature of our executive, concern about future outcomes tends to weigh heavily and causes mental fatigue; its power reduces as we spend large amounts of energy considering all the possible options. The fatiguing effect of the unknown is in fact greater than knowing for sure that there will be a negative outcome. In a study where participants were left unsure as to whether they would be giving a speech later that day (deliberately made more intimidating by being told that it would be rated by others), they were not able to concentrate. They made more errors in a task than a second group who were informed that they would definitely be in this anxiety-provoking situation later.[35] The depleting nature of this ambiguity on the executive was also shown by the first group's higher consumption of the available sweets – they were more likely to act with their autopilot, reaching for immediate rewards, something that even affected those who usually had good control over their eating habits.[36]

Stress eating is a concept most of us are familiar with, and are likely to have personal experience of, but what might surprise some is that our use of technology is very similar. This was highlighted in a study where participants were assigned a high-stress task. They were given five minutes to prepare a speech about why they were the ideal candidate for a specific job, which was to be recorded on camera, adding to the pressure they would have felt. These participants, who were surreptitiously observed in the waiting room during a respite in the task, ended up using their phones significantly more than those who were assigned a low-stress task – having to provide written advice for someone else who was about to start the job.[37] Just as stress can trigger overeating, this high-stress scenario led to increased phone use.

The pandemic provided an overwhelming amount of uncertainty and stress for masses of people. Its impact on our executive function was widely observed by many people who described suffering from 'brain fog'. While brain fog is not a medical diagnosis, its symptoms of reduced attentional capability, poor working memory and mental fatigue all point to an exhausted executive, fatigued from having to constantly adapt to ever-changing circumstances. Given our compromised executive function – taxed by uncertainty about the future and running in low power mode – we find ourselves increasingly dependent on our autopilot habits. When you factor in our strengthened associations to our phones, both within our homes and emotionally, and consider the limitations placed on our outdoor, non-technological outlets, it's unsurprising that the pandemic saw an increase in phone use.

Our devices became powerful links to connect us all and to enable us to continue working and studying, but in doing so our brain forged even more screen-time associations and reminders, overwriting a lot of the device-free activities that we were no longer able to partake in. These habits, having been coded into our neural circuitry, are not then easily switched off and have the potential to become problematic. While it is still too early to fully understand the long-term impact of the pandemic on our technology habits, some people report that new digital habits formed as a result of the pandemic have continued to persist even after restrictions were lifted.

6

Really Small Actions

In 2016, Instagram's unprecedented success was causing the company a multitude of problems. The initial premise of the app was to be able to share beautiful pictures instantly – captured in the moment. Available only as a mobile app, it made the process of taking and sharing pictures directly from your phone extremely simple – great for building a regular habit of uploading and interacting with images. However, as Instagram's user numbers skyrocketed, there was increasing pressure to post the perfect image. Rejecting the spontaneity of the founders' original vision, people were curating their Instagram feeds down to each exacting detail.[38]

Instagram was concerned; users felt each picture they posted had to represent a momentous occasion, good enough to take its place on the permanent record of their feed. Uploading a picture, something that had been automatic – a 'really small action' – became carefully deliberated. In their quest for perfection, people were posting less, with the average user posting roughly once a week.[39] Less posting meant less content which in turn meant fewer people were opening the app to see what was new. Habits were shifting, and with many competitors vying for people's attention, the space left by Instagram would not remain unclaimed. They needed a feature which would make uploading pictures even easier. They needed to make it a Really Small Action again.

Understanding the scale of habits

Whether we go to university, move house or apply for that job, the large actions in our life require a careful process of planning and deliberating. As we've discussed, these complex decisions are all calculated by the executive brain, which delegates the really small stuff to the autopilot brain. The simplistic autopilot brain is not able to encode complex actions and to be performed automatically – habits need to be small. And when I say small, I mean really small. The habits stored in the autopilot brain are always much smaller than people expect. Washing our hands after using the bathroom is a habit that is very much ingrained in most adults, having been taught from childhood, but the habit is not just the process of washing our hands. Each movement within our hand-washing technique is itself a habit, from which hand we use to turn the tap on, to how much soap we use, to the process of rubbing our hands together in our own individual sequence. This is evident in health professionals whose training includes considerable time and effort undoing some of those ingrained habits to be able to wash their hands in a more thorough and clinical manner.

The automatic way in which these really small habits are performed means that they can contradict the overarching goals that the executive has set. Here is an example from my own life. Being a doctor during the pandemic meant I was washing my hands even more than normal and this led to a case of dermatitis where I was starting to suffer red, peeling skin on the back of my hands. The solution for this is an antibacterial cleanser which is less irritating to the skin compared to typical soaps, thus helping to preserve the skin's natural oils and moisture. Yet, each time without fail, my autopilot brain would reach for the familiar pump of my regular soap rather than the cream-based cleanser adjacent to it. I would only realize my mistake when I felt an unexpected sting in my hands. It seemed an easy change to make, had I been able to catch myself in the moment and apply conscious thought but, a lot of the time, my autopilot brain was keen on repeating my stored habit activated by the external reminders that we learned about in the previous chapter.

Even seemingly complex sets of actions are actually formed from really small habitual building blocks. For instance, cooking our favourite meal won't be a single habit but will consist of dozens of habits such as how we get cooking utensils out of the cupboard, how we chop the vegetables and how we stir the pot. There may be some periodic inputs from the executive where we pause to check the recipe and consider what to do next but, once that's decided, the executive hands the reins back to the autopilot leaving us to think about something else.

Technology relies on really small actions

Technology is generally created with the aim of making our lives easier, to turn previously effortful actions into effortless ones. A remote control makes it easy to switch on the TV. Online shopping makes finding what you want to buy easier. Email reduces obstacles to communication. Contactless payments are faster. These innovations are undoubtedly beneficial but, being prone to following the path of least resistance, they can also be habit-forming. Keeping our phone within reach means that minimal movement is needed to access it. The easier the action, the less thought it needs. There is less input from the executive, making it easier to act on autopilot mode.

Prior to the Phone Age, publishing our thoughts took substantial effort and delay. Preparing books and articles generally requires many hours of drafting, writing and editing. This started to change at the turn of the century with the explosion of blogs where anyone could write what they wanted and publish it instantaneously online. While this was easier than getting an article published in a magazine or a book, writing a blog still required a substantial amount of conscious effort to manage a website or platform, therefore limiting habit formation. Social media created an alternative. It lowered the stakes of what it took to publish by providing a platform through which, after a relatively straightforward process of creating an account, sharing our thoughts or images became a really small action. Write a few words or snap a picture and, in a few steps, the whole world could potentially see it.

In the technology sphere, where success hinges on habit formation, really small actions are crucial. While Instagram had inadvertently raised the bar of what was deemed good enough for its feed, another app lowered it. Snapchat's concept started off as a rejection of the polished world that Instagram had created.[40] Its app embraced posting unfiltered, spontaneous, in-the-moment content called Stories which would then disappear. Compared to Instagram's highly curated world of 'really big actions' only done weekly, Snapchat was all about really small actions that could be done spontaneously, multiple times per day without laborious editing and planning. As a result, people created habits of chronicling their entire day: wake up – post – have morning coffee – post – finish a workout – post – to share that achievement. The really small actions provided by Stories had enormous habit-forming power.

To stay relevant, and habit forming, Instagram had to take drastic action. In August 2016, it launched its own version of Stories, a direct copy of Snapchat.[41] The bar was again lowered. Whereas Instagram feeds remained highly curated, a really small action was all that was needed to upload something to Stories. There was also a culture shift. Rather than extraordinary moments, people could post their day-to-day activities, forming numerous habits not just for themselves but for the people who follow them and who frequently open the app to view those updates.

Instagram and Snapchat's use of Stories is one of many examples of how smaller technological actions can be encoded into our autopilot brain to become more automatic. Tech companies are constantly innovating on this front: with face recognition, we no longer need to press a single button to unlock our phone; we are permanently logged into all our apps, from email clients to social media to games. Designers will notice a behaviour and find solutions to make it easier. Noting that when binge-watching, people skipped credits and episode recaps, streaming services created a single button that allows viewers to do just that. Even small details like double-tapping an entire photo on Instagram (rather than pressing a small button) are designed for convenience. This ease, however, comes at the cost of creating many digital habits.

That habits are really small does not describe their size but, rather, the amount of mental effort that is required to execute them. Writing a long blog post requires a lot of mental effort but so does having to edit down thoughts to fit into a single 140-character tweet. Realizing this in 2017, Twitter doubled the number of characters of their posts, ensuring that people would have ample room to tweet their thoughts but without significantly raising the standard – and therefore mental effort required – to post or read longer prose. TikTok, an app made famous for one-minute-long videos, gave users the option of producing longer videos, reducing the amount of editing needed. Instagram followed suit whereas YouTube, a platform known for its long-form videos, gave users the option of shorter-form content which takes less effort both to make and to watch.

While technology companies continually aim to make our digital tasks more effortless, non-technological activities still demand that we make the same effort. And, as we form more digital habits, the balance between these types of activities shifts further towards the digital ones. This is because habits increase effortlessness from within by allowing our brain to dial down mental exertion and switch to autopilot. This becomes particularly apparent in low power mode when our autopilot takes over. In such states, we pick up our phone to check off Instagram stories rather than carry out a non-technological activity of equivalent duration, but which is not habitual, as our brain deems it more effortful without the support of our autopilot.

This highlights a key conflict at play when it comes to our digital habits. As discussed, when we are setting goals that we want to accomplish, we often set ourselves high standards that can be difficult to maintain – but this is in contrast to how technology is constantly finding ways to make their products more user friendly. To correct this mismatch, it is important that you have a Plan B (Building Block 2) in place for when you cannot meet those high standards, such as when you're in low power mode. This technique, especially when coupled with Inserting Hurdles (Building Block 3) to increase the size of the technological actions that you want to avoid, helps reset this balance and stops your autopilot from automatically deviating to the easiest option.

The domino effect

If you were to observe a neurologist examining a patient, you would witness a complex sequence of movements that test the function of multiple nerves in the face, arms and legs. This is followed by tapping in strategic locations on the body with a tendon hammer to elicit reflexes. For a novice, learning this sequence can be daunting, but for a seasoned clinician, it feels natural. The difference is habits.

Small physical actions can themselves act as reminders to trigger subsequent actions. This means that a series of really small actions can link together, forming a chain, each act initiating the next like a falling row of dominoes. If you were to ask me to complete my neurological examination out of its usual order, I would need to use my executive brain to remember what I had already tested and what needed to be done next. However, the predictable chain of really small actions means that once I start and the first domino falls, the rest will naturally follow. This sequence can be executed by my autopilot, freeing my executive to focus on finding abnormalities and considering potential diagnoses, rather than what I need to do next.

This 'domino effect' can be observed in our use of technology. When you pick up your phone, you perform a series of actions (such as checking messages, email, news apps and social media) in a similar order each time, and each of us has developed our own unique app-checking cycle based on our stored habits. With close observation you'll find that, even within each app, the types of actions you do are similar. For social media apps, you may start by checking for notifications, scanning a feed or accessing other features, and may end with a final refresh before exiting to the next app.

Digital Distraction Cycle: Each Really Small Action can become a reminder to prompt the next action, like a falling row of dominoes. Sometimes, at the end of this sequence, enough time has passed that we feel the need to check the apps again, hence repeating the cycle.

A simple action such as picking up our phone to check the time may be the first domino to fall. Each subsequent individual action is really small but is able to trigger the next falling domino until they all add up to create a powerful digital cycle which is stored in the autopilot brain and is performed automatically. This is how the domino effect has the potential to turn each 'quick check' into an unintentional digital detour. In the London School of Economics study, this happened around 20 per cent of the time: four out of five phone checks remained as a single interaction but one fifth of the time participants were sucked into the virtual world and caught up in their very own personal digital cycle.[19]

Once such a cycle takes hold, it can be tough to interrupt. This is in part because of the lack of stopping cues that could jolt our executive brain into making a decision, and in part because of a sense of incompletion until all the actions have been carried out, in a similar manner to wanting to finish playing a frequently rehearsed musical piece. While our original intention may have been to check something quickly, before we know it, a considerable amount of time has passed. The duration of the digital cycle is difficult to control and its frictionless nature means that our brain often miscalculates the passing of time while we are absorbed by

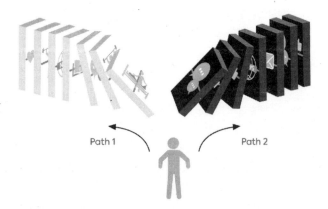

Domino Habits: How a slightly different action can produce dramatically different results via the domino effect.

our phones. To disrupt the cycle, we need to identify its components and Insert Hurdles (Building Block 3) at various key points. Hurdles halt the autopilot mode that the cycle is running on and activate the executive, giving you the power to make a different choice. When exhausted and in low power mode, this alternative choice can be small, such as delaying and using the Five-minute Rule (Building Block 1).

The impact of really small actions

By this point of the book, a large number of readers will have reached, or will have wanted to reach, for their phones. And it doesn't matter if you do. After all, it is just a quick check, and you can return to reading. If I were to pause writing this book and scroll on my phone for five minutes, nothing would happen. I could return to writing with no one being any the wiser. This is because the momentary impact of each really small action always seems negligible.

The conscious part of our brain, the executive, spends such a considerable amount of effort and energy thinking about the big stuff that it is easy to dismiss the small stuff that it has delegated to our subconscious autopilot brain. This means that we consistently underestimate the power of really small actions and overestimate the power of large actions. The executive brain will set out the overall vision but the legwork has to be done through habits that support our goals in the autopilot brain. Large, ambitious actions are few whereas habits are numerous. Our lives are not the result of a few giant decisions but rather the product of many millions of really small actions. For example, we will only move house a certain number of times in our life, but we will use things that need to be put away in the correct place countless times to live in a tidy home. This principle is particularly pertinent to our phone habits where, rather than relying on a one-time overhaul, we need to apply consistent effort to build and maintain supportive digital habits that will govern how we spend our time, focus our energy or nurture our relationships in the long term. Time also multiplies the power of a really small action. What seems like a minor five minutes each day adds up to thirty hours per year. This is practically the equivalent of a working week when you consider breaks, meetings and interruptions and is certainly not an amount of time that we would easily disregard. In the physical world, it is common to dismiss those five minutes, thinking that it is not a long enough period to do anything significant other than a quick browse on our phone: not enough time to read, but you can use those few minutes to scroll; not enough time to journal but you can tweet; not enough time to exercise, but you can watch a few TikTok videos instead. In the digital world, however, most technology companies are perfectly happy to be the recipients of those really small actions – those five minutes – that we dismiss as inconsequential. They know that those minute-long checks of their app, when taken together and given time, will add up to something significant.

As we learned in the previous chapter, the probability of doing any given action over another can depend on our environment and our internal state, but what we need to understand now is that it can also depend on what we did previously. Even within the same context, executing a really small action has the ability to completely alter our course through the domino effect. People often report that making their bed as soon as they get up leads to a productive day whereas staying in their pyjamas leads to the opposite. A little tap to push over the first domino and the rest follow suit. For one person, picking up their instrument will be the first domino to fall, eventually leading to a successful practice session. For someone else, putting on running shoes will be the first domino to going for a run. These initiating habits provide the launch point, without which the activity will not take place. In the same way that sitting at our desk or putting on running shoes puts us in the best position to write/work out, picking up our phone puts us in the best position to scroll and follow a digital distraction cycle. The domino effect has the potential to amplify the power of a really small action, whether it supports our goals or contradicts them.*

It is important to emphasize that, rather than being afraid to use technology, what we need to do is reclaim some of the balance between our beneficial and problematic digital habits. Our phones provide countless valid reasons for use, both for productivity and for entertainment, and it is perfectly reasonable to use them some of the time. But we must be mindful of the impact those really small actions are having and the digital habits we are creating in our autopilot brain. The innocuous nature of our habits and their ability to go unnoticed means that they invade parts of our life undetected, multiplying through time and through the domino effect. So, as I am writing this, I could quickly check my phone for five minutes and then get back to writing and no one would know. But how would I feel if this was to happen multiple times a day for an entire year, slowly sucking me in, diverting my focus in such a way that I have trouble finishing my manuscript? I know my answer. It's now time for yours.

---

* While writing this book, my domino habit was to write a single word. Sounds silly but it was very effective. This still meant I got a cup of coffee, sat at my desk, opened my Word document and read what I last wrote before deciding what to write next. A cascade would often follow. When my life got busy, it would have been easy for weeks to pass with no writing. My single word per day habit meant I could build a powerful writing habit in my brain rather than letting weeks go by without doing anything, as so often happens when life gets busy.

7

Reward

Despite its role in shaping a society that thrives on digital currencies, the creation of Facebook's 'like' button was, in fact, something of an accident.[42] Its original purpose was to streamline laborious responses to posts and replace them with a really small action. For example, multiple congratulatory comments following an important announcement could be replaced with the simple tap of a button and consolidated into a single number. This really small action meant that engagement with posts would increase, no further typing necessary. People would still be free to leave a longer, more heartfelt comment which, instead of being lost among similar responses, would stand out. The 'like' button achieved its intended purpose but it also had a rather unintended, rewarding consequence. People started posting more in order to get likes.

We instinctively know that our phones are rewarding. We may reach for them to provide entertainment or interest, to satiate our human need to communicate and engage with others, or to use them as a distraction – a handy alternative to a moment of boredom, or to a difficult task that we want to avoid. But what happens in our brain during those moments? So far, we've learnt about the two core components of our brain machinery, and we've talked about how reminders in our environment and our internal state act as the first piece of the Habit Puzzle and become linked to the second piece, our really small habitual actions. In this chapter, we will focus on reward, the third piece of the Habit Puzzle. It works together with repetition to bind the really small habitual actions to their triggering reminders. To explain how this happens, we have to delve deeper and go even smaller. We have to talk about a chemical found in the brain – dopamine.

Reward saves habits

Just beneath the autopilot but above the brainstem, where the brain matter begins to narrow and form a stalk that attaches to the spinal cord, is the midbrain. Here, a small group of neurons (roughly less than 1 per cent of the total number in our brain) form the ventral tegmental area. It is one of two major dopamine regions of the brain, the other being the substantia nigra which is situated within the autopilot itself, its name derived from the Latin word for 'black substance' owing to its characteristic dark appearance. Dopamine is one of the main neurotransmitters found in the brain, a chemical that our neurons use to communicate with each other. Our brains function through a steady background level of dopamine, which gently ebbs and flows. However, it can also be released in rapid-fire bursts that occur on a millisecond scale. The baseline levels of dopamine and the timing of its rapid-fire bursts, followed by the subsequent brief pauses, create a kind of dopamine Morse code that scientists have been trying to decode for over seven decades, yet we're still discovering more about its complexities and functions.

If you were to do a quick search on the Internet, you would most commonly see dopamine referred to as 'the pleasure molecule' but the role of dopamine in the brain is much more complex than that. For example, dopamine produced by the substantia nigra has a key role in movement. This function can be most commonly observed in Parkinson's disease, a condition where the dopamine cells in the substantia nigra become damaged, and which results in slow movements and tremor. While the substantia nigra has a key role in movement, the dopamine cells in the ventral tegmental area have a key role in reward signalling. Rather than thinking of the popularized and simplistic view of dopamine as the pleasure molecule, I want you to think about dopamine in this region as having the following two functions:

- It provides a learning signal
- It motivates us to act

The ventral tegmental area has large, motorway-like connections direct to the autopilot brain. Dopamine released from here when we receive a reward indicates to the autopilot brain that it needs to assess the situation. It needs to look back. What was the really small action that led to this reward? What were the surrounding reminders in the environment? The dopamine released during a reward therefore provides our brain with a powerful learning signal and starts to change our brain connections in such a way that a link is constructed between the reminder and the really small action. In essence, reward begins the habit-saving process and repetition finishes it. This saving process updates the habit sequences stored in our autopilot brain. When the autopilot brain stumbles upon the same circumstance, it will seek to repeat the same rewarded action.

Our brain is constantly assessing the world and making predictions of what will happen. It therefore needs to know when these predictions are wrong. This means that unexpected rewards release more dopamine and provide a bigger learning signal than expected rewards. The exact value of a reward is less important to our chemical signalling than the reward being unexpected. For example, winning some money will give someone a bigger boost even if the amount of money is small when compared to their overall salary. This is because our salary is predictable whereas the extra money is a surprise. Guaranteed rewards do not excite our brain, in the same way that we would find a gameshow where all contestants were guaranteed the same win every time unwatchable.

Reward therefore plays a crucial role in encoding digital habits into the autopilot brain. Take the situation I've discussed already, of sitting at your desk to begin work. You may find yourself unable to concentrate, something not unexpected if you are in low power mode. Looking for something else to do, you may open your social media app. You post something. A familiar ping follows. There are notifications. Lots of likes flood in. Your boring morning has just been made more exciting. There is a spike of dopamine. The captivating content on your phone has provided a learning signal for your autopilot to associate this really small digital action with reminders in your external environment (i.e. your desk) and your internal state (i.e. being bored). It is the beginning of a new digital habit.

Prior to the introduction of Facebook's like button, the extra effort needed to type a comment meant they were relatively few and it was impossible to know who had seen your post (if anyone at all). The sparseness of comments naturally limited their rewarding nature, but the like button changed that. The really small action of liking a post meant more people did it. The reward of social validation suddenly had a number, and checking notifications to see an unexpected flurry of likes come in was powerfully rewarding. More than that, it was a learning experience for the brain. The habit was saved. The number of posts increased as a result.

Whereas a large part of our daily routine is monotonous, and our brain knows what to expect, the vastness and breadth of the digital world means that there is greater potential for unexpected rewards. We are more likely to stumble on an unexpected reward by looking in our phone than in our physical environment, such as a positive email received out of the blue, a particularly captivating news story or a message from a friend. These unanticipated rewards provide a bigger dopamine learning signal. The digital world is also able to provide rewards for aspects of our lives that would not ordinarily be rewarded. A random distracting thought that would otherwise quickly be forgotten can be rewarded by following a train of interesting links on the Internet. The same fleeting thought when posted on social media might garner likes.

Phones not only reward us in what they enable us to do but in what they enable us to avoid. A quick check of our phone can provide a reward in the form of a temporary reprieve from a difficult task, a moment of boredom or an awkward social situation. Part of the rewarding nature of our phones is that there is no hard work attached to them. Unlike activities such as studying, exercising or doing household chores, our phones offer easy access to information and entertainment without any significant effort required.

The motivation molecule

Dopamine wields a dual power – it not only aids learning but also acts as a robust motivator. Its latter role was highlighted in a series of groundbreaking experiments conducted in the 1990s involving monkeys and fruit juice. Just like us, monkeys think fruit juice is tasty, and the prospect of obtaining some incites considerable excitement, leading to a detectable spike in their dopamine levels. During these experiments scientists would consistently switch on a light, a few minutes before bringing out the fruit juice. At first, the monkeys paid no attention to the light, failing to associate it with the forthcoming treat. However, dopamine's role in learning prompted a shift in their behaviour. Their brains began to actively associate cues in their environment – like the light – with the rewarding outcome of the fruit juice. Soon, the mere act of switching on the light sparked anticipation in the monkeys, evidenced by lip-smacking even before the juice appeared. What was even more fascinating was the shift in the dopamine signal itself. The surge in dopamine, initially aligning with the reward, was now triggered by the light – the reminder. Dopamine was being released in anticipation of the reward, rather than in response to the reward itself.[43]

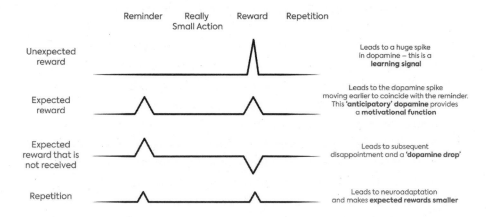

Dopamine Dynamics: How dopamine levels change in four key scenarios. Contrary to its reputation as the 'reward molecule', dopamine plays a crucial role in learning and motivation.

This experiment beautifully illuminates dopamine's dual role in anticipation/motivation and learning. When the reward is unexpected, dopamine will provide a learning signal, triggering the autopilot brain to look backwards at what preceded this reward. However, once the brain learns to accurately anticipate the reward, it starts to look forwards and anticipates the reward instead. This surge of dopamine is what motivates us to get the coveted reward. In scientific terms, dopamine provides an estimate of how worthwhile it is for us to expend our finite energy resources to obtain a particular reward. The bigger the reward on offer, the more intense the dopamine surge, and the stronger our drive to act becomes.

This is what makes digital habits so difficult to manage. Unlike the single reminder light that the monkeys had, we have many reminders, both through our external environment and through our internal state. Our bed, our desk, the dining table and even the bathroom (for some) have all become reminders signalling potential rewards should we reach for our devices. Our internal state – whether we are feeling happy or sad, bored or anxious – will also do the same. Our brain, on encountering such reminders, will be anticipating a digital reward on a subconscious level even before we've touched our phone. Not doing so requires the executive brain to step in and exert precious mental energy to override the autopilot. The fatiguing effect of doing so multiple times per day is more likely to put us in low power mode. Unfortunately, in that state, we are even more susceptible to the immediate rewards readily provided by our phones.

This anticipatory dopamine spike means we are willing to check our email and social media feeds tapping our screens multiple times, even if most of what we encounter is dull. Our brain decides that this action is worth it for the sheer hope that there will be one interesting notification, one possible reward. It is the alluring promise of a potential reward that fuels our behaviour rather than the reward itself. We 'want' to see if something exciting has happened and this desire persists, despite the fact that we often don't receive the promised reward – for instance, our tenth social media check in the same hour is unlikely to offer any additional gratification. But 'wanting' the reward is very different from 'liking' the outcome. We repeat this behaviour over and over again in the hope that the reward will appear despite ending up feeling bad about ourselves after we've been distracted one more time.

Reward by design

Although the creation of the Facebook like button was reportedly accidental, features of our phones and apps are deliberately designed to increase reward. These design choices are not limited to technology but are present in a variety of sectors, all of which use neuroscience insights to increase habit formation. How these are implemented may differ, but they are all around us. For instance, when dining at a restaurant, you may be given an unexpected free dessert at the end of your meal. When shopping in a supermarket, you may receive some money off vouchers for your next shop. When walking into a coffee shop, you may find a seasonal variation of your regular latte. These unforeseen rewards are strategically placed to try and press the save button on your habits and motivate you to continue them.

An illustrative example of how rewarding design influences people's behaviour can be observed in the algorithmic feeds of social media platforms. The initial excitement and novelty of new social media platforms was inherently rewarding. However, as the user base grew, they began to lose their appeal. At first, social media feeds displayed posts in a straightforward reverse chronological order based solely on posting time: the more recent posts would always be on the top of the feed. However, over time, this approach made it harder to stumble upon truly rewarding content. Instead, the feeds became dominated by high-frequency but low-quality posts, like blurry pictures of dinners, burying more exciting announcements or joyous news amid the noise. While anticipatory dopamine allows us to sustain habits without constant rewards, it is crucial to maintain a balance. In the case of social media usage, the habit can only be sustained if rewards are frequent enough to capture our interest. Without engaging content to captivate us each time we open an app, disengagement becomes likely. If the rewards become too infrequent, our brains will seek out other, more rewarding digital activities – in practice this means switching to an entirely different app. Ensuring that rewards are frequent enough to maintain our engagement and interest is therefore vital to sustaining a digital habit.

To enhance the frequency of rewarding content, social media platforms generally implement algorithms that prioritize posts based on user engagement. This process is akin to a newspaper featuring captivating headlines on its front page. However, in the case of social media, it is not human editors making the selections but sophisticated computer code. The algorithm behind your social media feed not only prioritizes content that receives collective engagement but also customizes recommendations according to your personal preferences. By harnessing data from your online interactions, such as likes, comments, and shares, the algorithm also continuously improves its own predictions – this self-improvement is termed machine learning. This fine-tunes the ability to provide content that you will find rewarding, leading to increased habit formation. In line with that, Adam Mosseri, the current CEO of Instagram, openly acknowledges that internal testing shows that users find the older, chronological algorithm less rewarding, resulting in less time spent on the platform.[44]

Unlike other social media platforms, where you had to invest time and effort to find people to follow and populate your feed, a key feature differentiating TikTok, which was launched in 2016, was that it could deliver unexpected rewards with very little effort. As soon as someone signs up to TikTok, their feed is immediately populated with the most enthralling videos on the platform and, as the algorithm collects information about their likes and dislikes, it can personalize this feed further. When TikTok launched, its algorithm not only made watching videos on the app more rewarding but it made posting them more rewarding too. In contrast to the predictable likes delivered by other social media sites, posts on TikTok have a greater potential of going viral and providing an unanticipated reward. This is like hitting the jackpot in comparison to your expected salary and played a significant role in the platform's rapid rise in popularity.

The success of social media in embedding habits exemplifies the profound impact of unexpected rewards, further amplified by our inherent need for social validation and a sense of belonging. A flood of likes, a sudden surge in followers, or an unforeseen positive comment can trigger the urge to repeatedly refresh our phone screens. It is not uncommon for a single unexpectedly rewarding experience, such as going viral, to become the catalyst for habit formation. We become engrossed in closely monitoring our social media metrics, eagerly anticipating the next exciting occurrence and another rewarding moment to unfold.

Our phones are not drugs

It's crucial not to oversimplify and label dopamine as 'bad', as if it were some sort of irresistible yet perilous drug. Dopamine is an integral part of our neurochemical make-up and plays multiple vital roles, none of which is inherently negative. It's understanding and managing these mechanisms that's a real game-changer. Dopamine doesn't judge – it motivates us to seek rewards and repeat actions, shaping both supportive and problematic habits. However, its misleading tag as the 'pleasure molecule' has persisted, fostering the notion that our phones provide us with some sort of 'dopamine hit'. This unhelpfully emphasizes the drug paraphernalia narrative that is so commonly used to scaremonger our behaviours around technology.

Having dedicated a significant portion of my PhD to studying dopamine, I've endeavoured to provide a thorough explanation of dopaminergic signalling in this book, showing you how these concepts apply to our phone habits while being careful not to over-interpret the data. It is important to keep in mind that much of what's portrayed about dopamine in the popular media is typically overly simplified or purely speculative. In studies involving monkeys and fruit juice, for example, researchers measured dopamine level spikes using a method called microdialysis. This technique entails implanting a small probe into the brain to sample the fluid bathing nerve cells, hence enabling dopamine levels to be measured. However, this procedure, along with more advanced techniques that measure dopamine release from neurons, is far too invasive to use in humans. As a result, we glean much of our understanding about how dopamine functions from experiments conducted on animals, such as monkeys, mice, or rats. While these studies offer invaluable insights, they do have limitations, particularly when we consider how humans uniquely interact with technology.

We can use scans to image the dopaminergic system in humans, but these scans can only be done over extended periods of time. They serve a crucial role in diagnosing conditions tied to reduced dopamine levels – for instance, I might request these scans for a patient if the diagnosis of Parkinson's disease is unclear. However, they aren't capable of highlighting those individual, rapid bursts of dopamine that occur when we encounter a rewarding experience, such as checking our phone. Recognizing this limitation is vital when we try to translate lab findings to our everyday technology habits and attempt to understand our interactions with the digital world.

In a modern world where countless experiences are labelled as a 'dopamine hit', it's essential to remember that being rewarding is not synonymous with being harmful. Indeed, dopamine's role in learning and motivation is not just necessary – it's vital for our very survival. This becomes starkly apparent when we consider the struggles of those with Parkinson's disease. In these patients, low levels of dopamine result in not only difficulties with movement but also a significant lack of motivation. This state, known as apathy, means that even high rewards no longer provide the same motivational incentive that is mediated by dopamine.[45] For some of my patients this is severe enough that they no longer have the neurochemical dopaminergic drive to initiate activities that they enjoy – both digital and non-digital. The same lack of dopamine also means that someone with Parkinson's disease will have trouble with habit formation.[46] This fundamental role of dopamine becomes even more evident when we look at the world of scientific research. Consider, for example, mice that have been genetically engineered so that they cannot produce dopamine. These mice display a severely reduced desire to move, even when the temptation of tasty food is on offer. If left alone, they would quite literally starve. Without dopamine, we humans would face a similar fate. However, when the dopamine levels of the deficient mice are restored using L-DOPA – the same drug used in the treatment of Parkinson's disease – they spring to life, dashing about their cages and eating.[47] This powerfully illustrates the indispensable role of dopamine's motivational drive in our survival. It is this same dopaminergic drive that fuels humanity's endless quest to explore, to reach out for the new, and to constantly innovate. So, let's respect the complex role of dopamine in our lives – it's not a villain, but a vital part of what drives us forward.

Our bodies are all about balance. Too much of anything, even something like dopamine that our brain naturally produces, can lead to problems. But here is where we need to make a clear distinction between the dopamine released by our own brain's machinery and that of drugs. The dopamine surge triggered by our phones is the same chemical signalling we experience when we enjoy delicious food, see a loved one or pet an animal. Many recreational drugs are able to trick our brain into releasing more dopamine than is physiologically possible. This may contribute to the feeling of euphoria, and it's this artificial high that can lead to addiction. As we've seen in this chapter, dopamine will press the 'save' button on our habits but supraphysiological dopamine release through drugs will powerfully rewire brain circuits. This rewiring is the equivalent of turning down the volume on the executive brain – as we have seen, this is the part that assesses long-term consequences. Drug addiction changes the brain, reducing its ability to consider the harmful impact of drug use – on health, finances, relationships – and, with the executive being the source of self-control, someone who is addicted will struggle to resist.[48] Obtaining and using the drug gets prioritized, drowning out all else. In my practice, whenever colleagues have felt frustrated about a patient like Keira relapsing, I've explained this process to them. Everyday items like a spoon or a lighter, innocuous objects for most of us, have been so powerfully linked to drug use that they've become reminders potent enough to trigger a relapse. And that's what makes addiction incredibly challenging – it's not just about the substance. Those battling drug addiction are up against the powerful imprints it leaves on their brains.

Even medical methods which artificially increase dopamine levels can have adverse effects. Giving L-DOPA, the drug used to boost dopamine in the dopamine-deficient mice, to people with normal levels of dopamine impacts their decision-making, leading them to make more impulsive choices and become focused on short-term rewards, as a result taking more risks.[49] Neurologists, including myself, take great care prescribing a certain class of dopamine drugs, called dopamine receptor agonists, to patients with Parkinson's disease, warning them of the risk of developing an impulse control disorder which can lead them to excessive or pathological behaviours surrounding shopping, gambling, overeating or watching pornography.[50] Patients on these drugs require careful monitoring. Our technology is designed to maximize rewards and to be habit forming, but it is not an addictive substance, and that is what makes it possible to unwind our own phone habits without the interventions that are required in addiction.

Delay discounting

If we were to consider our phone use rationally, we would realize that the reward we get from checking our phone – for the hundredth time the very same day – is much less than achieving a future dream. For me, writing this book was much more fulfilling than a quick scroll on Instagram and I am sure that there are many things that you would find more satisfying also. If that's the case, why do we seemingly prioritize reaching for our phone so often? It's because the rewards provided by our phones have one significant feature – the timescale in which they occur.

There is a feature of our brain called delay discounting, where our brain automatically discounts the value of distant rewards. The further away the reward, the more its value is reduced. The anticipatory dopamine spike becomes smaller through this discounting effect, so that the prospect of winning a large sum of money tomorrow might have you daydreaming about all the things you could do with the money, but this giddiness reduces when thinking about getting the same reward a year later and even more so when considering the same reward in ten or twenty-five years' time. Anticipation of a future reward tends to build when it is in close proximity.

Delay discounting is a survival mechanism to ensure we look after our immediate well-being. This may be useful in some situations but, for most of us, there is no immediate danger and our brain is overcompensating. Delay discounting means we still value future rewards but that they have to be larger to make up for the discounting that is implemented by our brain. This is where technology has a disproportionate advantage. With instant publication of our thoughts, quick delivery times and not needing to wait between episodes, technology provides more powerful rewards due to their imminence. This is why the prospect of a smaller immediate reward such as checking our phone seems more irresistible in the moment, outweighing distant, but more important, accomplishments.

Delay discounting is the mechanism that makes Pre-commitment (Building Block 4) so effective. Due to this tendency, we undervalue rewards that come in the future, making it easier to make healthier choices ahead of time than when weighing up our options in the present. For instance, when planning meals for the upcoming week, you're more likely to choose nutritious options than when you're deciding what to eat imminently, particularly if you're hungry. The same concept applies to our digital habits – the thought of checking an app tomorrow seems far less tempting than the almost irresistible urge in the present moment, especially when you're operating in low power mode. Even short delays can have a significant impact. If you have been applying the Five-minute Rule (Building Block 1) and waiting for those minutes to pass, you will have found that, in many instances, you no longer want to check your device – this short delay being enough to reduce the appeal of the reward.

The difficulty in building many of our desired supportive habits is that the long-term rewards they provide are often discounted. We can address this by hacking the system and changing the timescale in which we gain these returns, using short-term rewards to build habits that provide long-term gains. For example, in a study where participants only had access to tempting audiobooks at the gym (short-term reward), this increased their gym attendance (long-term reward), as the immediately rewarding nature of the audiobooks helped cement the habit.[51] Two activities do not need to be done simultaneously in order to be linked. Rewarding yourself after the habit is performed works well too. This does not mean that you need to create a new reward for each habit – inserting a supportive habit before you would ordinarily have a treat works just as well. For example, if you indulge in a beloved TV show every evening, you could use this as motivation to perform a supportive habit, like doing a quick cleaning sweep of your home beforehand. In this way, the joy of watching your favourite show becomes associated with the previously mundane task.

When thinking about rewards, most people consider external or physical rewards, but internal rewards can be just as powerful. Inner critical thoughts, such as thinking that you've not done enough or could do better can sabotage habit formation. Understanding that each really small action you take is a step towards habit formation, and is rewiring your brain, makes a big difference to how you think about your performance. Make sure you give yourself a reward – in the form of a metaphorical, or even a literal, pat on the back – every time you apply the rules in this book. This is just as important.

Regulating rewards

When evaluating technologically rewarding activities, we must remember that the issue doesn't lie in the total time spent but rather the degree of control. The crux of the problem is the amount of unintentional and unregulated time. This was clearly illustrated through a large study which examined binge-watching. This study involved over 4,000 participants divided by the researchers into four distinct groups, each with their own unique characteristics.[52] At one end of the spectrum were the 'avid binge-watchers', a group that derived immense satisfaction from their TV time, choosing to immerse themselves in a broad range of series. Despite spending substantial periods engrossed in various TV genres, they did not perceive their viewing habits as problematic, but rather as an intentional, enjoyable activity. In contrast, the 'recreational TV viewers' did not derive the same fulfilment from watching content and consequently, they engaged in it less frequently.

The third group, known as 'unregulated binge-watchers', is of particular interest. These individuals spent a similar amount of time watching content to the avid binge-watchers. However, whereas the avid binge-watchers were intentionally indulging in a true passion of theirs, the unregulated binge-watchers lacked this intentionality and purpose. As a result, they struggled to regulate their viewing time, often spending significant periods unintentionally binge-watching, and as a result, they found their viewing habits problematic. This is in direct contrast to the last group, the 'regulated binge-watchers', a group who were able to maintain control over their viewing habits – in doing so, they found satisfaction in their binge-watching while demonstrating moderation.

Upon reading this, you may have quickly pinpointed which group you fall into. This idea stretches beyond binge-watching to other activities like gaming, scrolling through social media, or online shopping. If you're an 'avid user' of technology, this is fuelling a passion of yours and this is how you have intentionally chosen to spend this time, that's perfectly fine. However, if you identify with the unregulated group, it's key to establish healthier digital habits to regain command over your tech use. By moving towards the regulated group, you can reduce potential problematic behaviours and regain your sense of control. You'll stop viewing technology as an issue, curb your frustration, and start to enjoy it in a more controlled, meaningful way.

Learning to practise anticipation is key in regulating rewards. It is a common misconception that the most pleasurable part of a reward is obtaining it. Since dopamine is released in expectation of a reward, there is actually greater value in anticipation. In essence, it's not just about receiving rewards, it's about having something to look forward to. This is seen in our daily lives, where our collective mood tends to rise throughout the week in anticipation of the weekend. Our mood generally peaks on Saturday when most of the weekend lies ahead. However, by Sunday, even though many still have a day off, our mood starts to dip.[53] This is because our forward-thinking brain is already contemplating the arrival of Monday. We count down the days until Christmas. We eagerly anticipate our next holiday and, once it's over, we start to plan the next one so that we have something to look forward to. Sometimes our anticipation of a reward is so powerful that actually obtaining the reward feels like a let-down. Variations in our genetic make-up and hence our dopamine signalling means that we all experience these feelings to a different extent, but this is something that really hits home for many of my readers who report that the run-up to a reward is often better than the reward itself.

With that in mind, let's take the concept of regulated vs unregulated binge-watchers a step further and paint a picture of these two groups' experiences of reward. Someone who is a regulated binge-watcher likely derives heightened pleasure by savouring the anticipation between episodes, thereby garnering richer satisfaction from fewer, intentionally timed rewards. Conversely, unregulated binge-watchers, driven by instant gratification, ironically would experience diminished enjoyment as a result of overuse – this is likely to result in a less rewarding viewing experience, compounded by guilt and frustration over their perceived problematic behaviour.

In our interactions with technology, we can often fall into the unregulated category, tantalized by the promise of reward. However, the imminently rewarding nature of tech means that we hardly ever get the time to look forward to anything anymore. We can obtain a large number of rewards on a very short time scale, sometimes even instantaneously. It also means that any rewards are fleeting – they are over so quickly that we get no build-up and there is no time to anticipate. In essence, we are surrounded by rewards but have less to look forward to. The solution to this is *not* to remove these rewards aiming to make our physical or technological environment more 'dull'. I am personally grateful that my social media feed is tailored to my preferences and that I have captivating shows to watch in the same way that I am to have tasty food to eat and engaging books to read. These rewards enrich our lives but we must regain some balance – this is a central part of my personal ethos.

The key to this balance is regulated reward, not complete deprivation. What this means is, rather than loading up on many instantaneous rewards, we can make the most of anticipatory dopamine by reducing the number of rewards and practising anticipating them more. It is the equivalent of 'quality over quantity'. This, together with understanding that repeatedly reaching for our phones provides diminishing returns in terms of reward, can curb our overuse and allow us to use them more intentionally. And with this in mind, we are ready to tackle the final piece of the Habit Puzzle – Repetition.

8

Repetition

Much of Henry's life was dominated by epilepsy. A result of head trauma due to being knocked down by a bicycle at the age of seven, his seizures progressively worsened until they became so frequent and severe that he was unable to leave his home. At the age of twenty-seven, despite high doses of medication, his condition continued to deteriorate, leaving him with limited treatment options. As a last resort, it was then recommended that he undergo experimental brain surgery. The effects of the brain surgery he underwent in 1951 made him one of the most well-known patients in neuroscience, known only by his initials, H.M. for over fifty years, until the time of his death where his full name – Henry Gustav Molaison – was released.

The surgery controlled his seizures, but when Henry woke up, there was a startling consequence. Limited knowledge of the inner workings of the brain at that time meant that the function of many areas of the brain was not fully understood. It soon became apparent that the two golf-ball-sized areas removed from Henry's brain contained the hippocampus, our memory inbox. We have one on each side and the hippocampus is the recipient of all our new memories before they are moved into permanent storage and distributed in other parts of the brain. So, while memories from his past were intact, the lack of a memory inbox meant he could not form any new memories. He could maintain a conversation but, as soon as his attention was diverted, the contents of what was spoken about would be quickly forgotten. Henry remained stuck in time, believing that he was in his twenties right up until the time he died, aged eighty-two.

However, while Henry could not remember new experiences, his brain could still learn. An unexpected discovery was that, when asked to trace a star on a piece of paper guided only by its reflection in a mirror (something that is visually counterintuitive to all of us), he progressively got better and better. His amnesia, however, meant that he never had a single memory of practising. He approached the task as if it was new each time, carefully listening to instructions, only to then be pleasantly surprised by his better-than-expected performance.[54]

It turns out that, despite his profound amnesia, he could still create habits. This is an intriguing contrast to conditions like Parkinson's disease which damages the brain's dopaminergic machinery that is crucial for the third piece of the Habit Puzzle. Consequently, patients with Parkinson's find it challenging to establish new habits, even though they retain the ability to form new memories. Conversely, patients like Henry, despite having profound amnesia, will still develop habits due to their intact autopilot system. In research studies, patients with amnesia will outperform those with Parkinson's disease in tasks that hinge on the development of habits, despite having no conscious recollection of having practised or performed those tasks before.[55]

For Henry and for all of us, as habits rely on the subconscious part of the brain, they can be formed without us realizing. However, to be ingrained in the brain, they necessitate a key element. This is the final, pivotal piece of the Habit Puzzle: repetition.

Repetition builds habits

A single action provides little meaningful change. You cannot get fit from one workout or earn a degree from one study session. Repetition has the potential to amplify each and every action: laying a single brick is meaningless but repetition transforms it into something more tangible, a building. All our habits, whether digital or non-digital, have been built through repeated actions. So far, we've seen that the reminder provides the initiating cue for the autopilot brain to execute the really small action, the habit. Reward releases dopamine which starts to link the first two pieces of the Habit Puzzle, binding the reminder to the really small action to save the habit. But this is only the beginning, and unlike a single press of a button on a computer, saving a habit in the brain requires our final puzzle piece, repetition.

At the time of Henry's surgery, scientists thought that our brain was fixed with limited ability to change. It turns out they were wrong. Our brain is constantly changing, so much so that I can assure you that yours has changed already while reading this book, and it will be even more different by the time you've finished. We've learned that each of our brains has billions of neurons which can form an innumerable number of connections with each other, referred to as synapses. Those synapses are described as plastic due to their ability to constantly be remodelled, like plasticine, hence the term synaptic plasticity is used to describe the storage process through which our habits are saved. Connections in our brain grow stronger with use and weaken with disuse. As a result, we have moulded a brain that is a unique imprint of not only our personality but every one of our experiences and every moment of our daily life.

It is common to see a novice fumbling to do something that an expert will do seamlessly. We've all been there in one form or another. For most, learning to drive is initially hard. The multiple movements required to control a nearly two-tonne machine, applying just the right amount of force on the pedals, changing gears and turning the steering wheel, require a great deal of concentration from a beginner but they eventually become second nature until we no longer think about them. We just drive. Repetition builds and strengthens the neural connections in our brain making it easy to do an action, as the neuroscience maxim that 'neurons that fire together, wire together' describes. This is similar to forming a path through a forest; with enough repetition, the path becomes well-worn and effortless to traverse. Once the path is established, what was once challenging becomes more automatic and needs less conscious effort.

Through this process, our brain changes. As we know, all actions initially rely on our executive brain and repetition trains the autopilot to do an action automatically – it becomes a habit. If you scan a person's brain after repeated practice, you will see increased activation of their autopilot.[56] The executive is relinquishing control while the autopilot takes over some or all of the strain. Being able to engage the autopilot brain as well as the executive is like having extra brain power which makes the action less mentally fatiguing. This a process that we are unaware of until we look back and remember how difficult it was to do something for the first time.

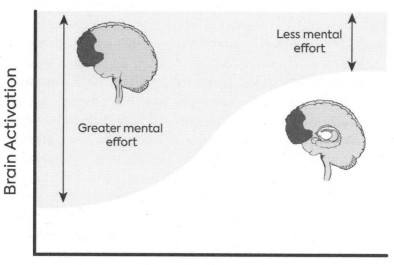

Repetition Increases Autopilot Activation: Initially, actions heavily rely on the executive, demanding more mental effort. With repetition, there is increased activation of the autopilot which supports the executive. This shift reduces the mental strain required to take those actions.

Technology encourages repetition

We instinctively know that the best way to create a habit is to do something frequently – on a daily or, at least, a weekly basis. Infrequent events seldom become habits as they are not repeated often. This is why brushing our teeth is a habit whereas renewing our car insurance is not. Of course, doing something more than once a day is even more likely to form a strong habit. With the average person reaching for their phone hundreds of times per day, this number is several magnitudes greater than almost any other task present in our daily lives, and it is that repetition in particular that builds up technological habits and makes them deeply ingrained.

Encouraging repetition is a tactic frequently employed by businesses to encourage consumers to develop habits around their products or services. A coffee shop may offer customers a free drink after a certain number of purchases to incentivize them to return. These types of loyalty programme aim to increase repetition and couple it with reward (as shown in the previous chapter) to foster habit formation. There are many ways in which our technology tries to encourage repetition. Some may be subtle, such as the ephemeral nature of Instagram and Snapchat stories which disappear after twenty-four hours. This impermanence provides an inconspicuous target of regularly posting or creates a sense of urgency, prompting us to regularly check the app to avoid missing out on updates. Other features more obviously enhance repetition; Snapchat, for example, provides a measure of how many consecutive days you have interacted with friends and maintaining this Snapstreak requires using the app on a daily basis.

Starting a new habit can be a frustrating process, especially when progress seems slow or non-existent, but technology overcomes this by rewarding repetition. Designers ensure that there are multiple ways in which we can monitor our progress. In addition to the Snapchat streak, we can move through levels in games as well as amass posts, followers and likes on social media sites. The metrics on most social networks are prominently displayed, signalling our progress through each post, their number highly visible on each user's profile on both Instagram and Twitter. One of the prominent metrics on TikTok is the cumulative total of likes obtained, a number that is certain to increase with each video posted. Showing us our total investment means that it becomes something we don't want to lose – and so we repeat.

Encoding new habits

There is a long-perpetuated myth that a habit takes twenty-one days to form but this is not based on science. One of the first, and most quoted, scientific studies found that the median time to form a new habit among all participants studied was sixty-six days[57] – three times as long as the twenty-one-day rule – and a similar number has been confirmed in later studies.[58] However, this does not mean that sixty-six days is the time that it takes to form all habits. What is less commonly reported about these studies is the very large range in which participants created a habit, varying from eighteen days all the way up to 254 days, the highest figure in fact mathematically extrapolated given that the study was only eighty-four days long.

How long it takes to form a habit depends on the individual habit and its corresponding Habit Puzzle. An easy action with a large number of reminders which is highly rewarded and repeated often will become a habit quickly. Research studies on habit formation primarily focus on health-related behaviours such as exercising, eating more fruit and vegetables or drinking water. Technological habits are much easier to build because they so efficiently fulfil each part of the Habit Puzzle. Checking your phone is a really small action with multiple reminders and which is highly rewarded. Rather than being repeated once per day, it is something that is repeated multiple times per hour and, as a result, habit formation is accelerated.

Instead of thinking about habit formation in days, it is best to think about it in terms of repetition. Behaviours that we do once per day will take longer to become a habit compared to checking our phone, something that is repeated multiple times per hour. Anything that we do only infrequently is unlikely to ever become a habit. As a general rule, the strength of the habit increases with the number of repetitions. This, however, is not a linear relationship as each individual repetition does not have an equal influence on the formation of the habit. Early repetitions have the most effect on building habit strength and determining whether the habit will stick or not. After a few repetitions, even though the habit may not be fully formed, the initial hurdle of starting will have been overcome and momentum will be in our favour. As we continue to accumulate repetitions, habit strength begins to plateau as it reaches a near maximal point. Habits that we have had for years have reached maximal strength with no further increase to be gained.

Therefore, getting started is the hardest part when it comes to forming habits, regardless of whether they are technological or non-technological. Moreover, the functions on our phone are not so irresistible that habit formation becomes inevitable. For example, many of us have downloaded apps to our phone with the intention of using them, but they've ended up cluttering our home screen and never being opened. Whether we form an app-based habit is decided early on, during the initial stages of repetition. There is a make-or-break point where the habit either gains strength or falls by the wayside. This was evident with the new Meta-owned social media platform, Threads. Despite the initial buzz and excitement drawing in 44 million sign-ups, its daily active users dropped by 70 per cent in the following weeks, suggesting that many hadn't established a strong habit of using the platform.[59] If we do start to use an app after downloading it, then habit strength will increase shortly after.[60]

App developers often try to increase the likelihood of repetition by prompting you to turn on reminders and notifications when an app is installed. These features are designed to increase the chances of getting over that make-or-break point in the early stages of habit formation. They are important when a habit path has not been built yet but, once the path is established, the autopilot brain is likely using a number of reminders in our external environment and internal state rather than solely relying on notifications. This make-or-break point early on holds true for most habits, not just digital ones. The main difficulty that people face when changing embedded habits is that they initially require motivation from the executive to reprogram them. Generally speaking, motivation is highest at the initial stages but depletes thereafter, particularly if we find ourselves in low power mode. For many, this motivation runs out before enough repetitions have been performed to insert the habit into the autopilot, and giving up at that stage means that no substantial change has been made. We then have to start again at the beginning.*

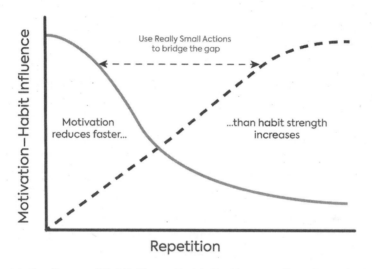

Motivation and Habit Strength: Motivation usually reduces before habits have had a chance to become established. Really Small Actions are a useful technique to bridge that gap.

* This will have a greater impact on those with executive function challenges, as they experience a quicker depletion of peak motivation. This in turn widens the gap between the fading of motivation and the successful encoding of new habits. In conditions characterized by dopaminergic disturbances, such as Parkinson's disease, or ADHD where dopaminergic dysregulation is a key focus of research, the brain's 'save button' for habit formation essentially becomes faulty. Strengths in other components of the habit puzzle can compensate for these weaknesses, so additional repetitions are required to establish a habit. However, this also has the effect of further widening the gap between motivation and habit formation.

Bridging this gap is important and we can apply the same lessons that technology designers use to develop supportive habits in our own lives. As we saw in Chapter 6, a big part of tech companies' approach is to simplify and make things easier, creating really small actions. Similarly, in our own lives, rather than berating ourselves to 'try harder', we should focus on making our actions small enough so that they can be easily repeated, especially during low power moments where we'd usually reach for our phone. Instead of unrealistic expectations, we need to build habits slowly, brick by brick or really small action by really small action. The Five-minute Rule (Building Block 1) and Plan B (Building Block 2) are both examples of these small actions that increase repetition, making habit formation more likely, even during low power mode and in cases where the executive is not functioning optimally. By developing these habits we can bridge the gap between our initial motivation running out and the point at which the habit is cemented.

Repetition changes reward

Our habits are built on the foundation of reward, which initiates the saving process, and repetition, which completes it. However, once there has been enough repetition to complete the saving process, the initial reward becomes less important as it is no longer needed to press that save button. Through repetition, an action which initially requires conscious control, hence relies on the executive, is delegated to the autopilot. As a general rule, the more automatic the action becomes, the less reward is required to continue it. A habit may even become so ingrained, so automatic, that we continue to enact it without the need for a reward at all. This happens because the neural pathways supporting the habit have been so frequently used and are so well established, they become effortless for our autopilot brain to navigate. This means we continue to repeat the behaviour simply because we've done it so often before.

In Chapter 5, 'Reminders', we discussed an experiment where participants were secretly observed while eating popcorn in a cinema environment versus a conference room. During that experiment, the researchers took it one step further by giving some participants fresh popcorn, popped an hour beforehand, and others stale popcorn that was made a week earlier. After the study, the participants were asked whether they liked the popcorn. Unsurprisingly, those who got stale popcorn did not. But, upon weighing the popcorn bags, the researchers noticed that those who had a strong habit of consuming popcorn in the cinema ate it anyway.[30]

Strong habits which have been repeated many times are characterized by reward insensitivity, where the rewards associated with that habit have become less important over time, sometimes even non-existent. This is why many strong problematic habits continue to persist despite their negative effects. Reward insensitivity can be advantageous when building supportive habits because initially we may need to consciously reward ourselves to establish them, but as the habit becomes stronger, that reward becomes less necessary. For example, parents may need to praise their children and use reward charts to establish a toothbrushing habit. However, as adults, we continue this habit without the need for a sticker chart. The stronger the habit, the more likely we are to continue doing it even in low power mode when we are tired and going straight to bed seems a more rewarding option.

Of course, the popcorn example is a bit extreme. If you were consistently served stale popcorn at the cinema, you would possibly either change cinema or opt for a different snack. Usually, the reduction in reward we experience is a bit more subtle than that. What typically happens is that repeatedly obtaining the same short-term reward causes its initial allure to fade over time as our brain establishes a new baseline. For example, a slice of cake is a treat when consumed infrequently. However, when consumed regularly, it becomes a habit. There is a degree of reward insensitivity, and what was previously a reward is replaced by a feeling of disappointment if we were to miss out on our usual sweet snack.

Thus, we continue the habit more to avoid this sense of loss than from a desire for a reward. Sometimes, frustrated that our treat is not as rewarding as it used to be, we may escalate our behaviour to gain more reward. When the slice of chocolate cake becomes part of our daily routine, why not put some cream on top as a little special treat? Until, of course, this too loses its rewarding appeal.

This concept accurately describes a significant number of unnecessary phone checks. Reports suggest that some people check their phones as frequently as '10 times each hour' or 'almost always'.[61] Upon reflection, one might wonder if this is because they find it rewarding, or, akin to the stale popcorn eaters, if they're merely following an autopilot script of reaching for their phones. More often than not, it's the latter, and this behaviour tends to breed frustration. The intervals between checks are typically too brief for anything worthwhile to have conceivably happened. Yet, the habit of frequently reaching for our phone, once established in our autopilot brain, creates a sense of disappointment when not followed. The novelty of checking the same apps repetitively might wane through overuse, but the habit endures to prevent the discomfort of not checking. Frustration arises from this reward insensitivity, which resets our brain's baseline. To reclaim the elusive rewarding feeling, this behaviour then escalates, evolving into an almost constant habit of checking. Unfortunately, this pattern leads to the reward becoming further and further diminished through overuse, akin to consuming multiple slices of chocolate cake at once where the taste becomes less satisfying with each subsequent slice.

Unlike the predictable rewards we get from things like popcorn or cake, the rewards we get from our phones and social connections in the online world remain unpredictable. This distinction is key, especially when we are experiencing reward insensitivity. As we've learned, dopamine functions more as a learning signal and a motivator rather than a pleasure molecule. This means that we're often enticed to check our phones due to the potential of an unexpected reward. Among the numerous checks we make on a daily basis, there might be one that reveals especially valuable information – a significant reward. This possibility means we retain motivation to continue the habit, even when the much-anticipated reward doesn't usually materialize or if, even when it does, it is not as rewarding as expected.

Completely shunning all rewarding aspects of technology is not the solution. Instead, we should aim to engage in mindful use of technology and strive for a few, intentional checks. Actively choosing to indulge in something, thus regulating our rewards and leveraging the power of dopaminergic anticipation, provides a more satisfying experience than numerous trivial checks. The latter leads to feeling unsatisfied, and it's this unfulfilling nature of unregulated phone checking that partly fuels overuse, trapping us in a cycle of constant insignificant rewards with little regard for what truly matters to us in the long term.

Whereas repetition causes short-term rewards to lose their initial novelty, it has the opposite effect on long-term rewards. Engaging in activities that have long-term benefits, such as learning a new skill, may be frustrating at first as they rely heavily on our executive function, but become easier and more enjoyable over time, as sequences become encoded into our autopilot. What is classed as a good long-term benefit will be specific to each of us but personal growth and self-improvement are key features to look out for. Both digital and non-digital activities fall into this category so you can utilize technology to provide long-term benefits. For example, instead of repeatedly checking the same app dozens of times, some of those checks could be used to take an online course, learn a new skill, listen to audiobooks, or read informative articles. By investing in activities that yield meaningful, long-term rewards, we can counteract the cycle of constant short-term reward-seeking. Do remember that this is not an all-or-nothing approach. Short-term rewards are a source of pleasure in life. It is just that they would generate more enjoyment if they were less frequent, more varied, and less at odds with your long-term goals. Striking a balance between short-term and long-term rewards is key to both maximizing the pleasures of life and achieving our long-term goals.

Undoing the Habit Puzzle

When the four pieces of the Habit Puzzle come together, a habit path is built and is progressively strengthened. Deeply ingrained phone habits mean that reaching for our phone has become the default option in our autopilot in many circumstances. Once those connections are strengthened, weakening them takes time. Removing something from your brain is hard. This may be evident in that you cannot instantly forget a painful memory no matter how hard you try because there is no magic delete button. It takes time for memories to fade and emotions to be processed, much like it takes time for a habit path that has been built to diminish and gradually disappear.

If you ever feel overwhelmed by the number of habits you need to change, remember that problematic habits tend to stick out more than the many supportive or neutral habits we have. Their negative nature means that we naturally focus on them more. However, it is crucial to avoid getting too fixated on changing only bad habits and neglecting the supportive and neutral ones. Focusing on building supportive habits is important because, to speed up the process of undoing a habit, what you need to do is give the autopilot an alternative option. You must build an alternative path hence replacing an old habit with a new one. In doing so, rather than undergoing the slow process of unpicking your connections, you are instead strengthening alternative ones.

Nourishing your new path while ignoring the old one means that, with time, the autopilot brain will start to take a new habit route before the old path has been fully deleted. For a substitution to take effect, you must replace your phone habit with an equivalent really small action which fulfils the same or similar needs as the phone habit itself. Take, for example, the fatigue-driven phone checks; the alternative activity you choose must also provide a comparable type of microbreak to the phone check itself. For instance, you might substitute one of your routine phone checks with a quick stretch, a walk or watering some plants. Avoid replacing it with an activity that you deem 'productive', as this won't meet the same need and will likely be unsuccessful. The Part II practical will delve deeper into how to reprogram your autopilot mode and offers an opportunity to strengthen and expand on the foundational habits of Part I.

Part II Practical

Reprogram Autopilot Mode

Now that you are equipped with the knowledge of how digital habits become ingrained, it is time to start changing them. This section consists of two steps: the first part guides you through how to break apart the Habit Puzzle formed by problematic digital habits, while the second part shows you how to form supportive habits in their place.

Step 1: Breaking Down Problematic Digital Habits

Building Block 5: Leverage Location

Reduce the number of location-based reminders to stop initiating habits.

- Location is a powerful reminder for our autopilot to trigger our habits. By leveraging location-based associations, you can pre-emptively prevent certain habits from becoming activated.

- Start building specific location habits. Set up rules for various phone activities or apps in alignment with these locations. If you have a problematic habit of checking the same app(s), designate 'digital zones' where you do this.

- Establishing these new, healthy patterns in a positive way is important: reframe locations as places you *can* use those functions on your phone rather than locations where your phone is not allowed.

- Simultaneously, harness the power of location to foster supportive habits. For example, designate your desk as a sanctuary for focus and your bedroom as a haven for relaxation. As time goes on, you'll find that instead of needing to fight off phone habits in these spaces, they simply won't be activated.

Some examples:

- Differentiate the areas in your living space which are for relaxation, focus and fun. The locations can be a specific room or rooms (such as the kitchen or the living room) or even a specific place within them (such as a particular chair).

- Are there functions on your phone that you would be better off checking on a computer or a tablet? Doing this automatically limits their location.

- Consider installing troublesome apps on an old phone and keeping that in one specific location.

- Keep your phone in a set location and use a smartwatch to ensure that you receive important notifications.

- For apps you find particularly problematic, you could set a complex password that is not easily memorable and place it in a designated location.

★ Keep in mind that the goal isn't to never check your phone; rather, it's about cultivating a more mindful and intentional relationship with your device, reining in those automatic, absent-minded checks. If you find yourself reaching for your phone on autopilot, don't feel bad about it. Simply relocate to one of your designated 'digital zones' as soon as you catch yourself. Over time, and with repetition, your brain will catch on and make the connection.

Redesign Your Digital Environment

Change the layout of your phone:

- Remove apps that are no longer needed. They create digital clutter that can be attention-sapping, as our brain needs to make the effort to ignore them while searching for apps that we use.

- Only apps that are either tools or provide a long-term benefit to you should be on the home screen.

- Move problematic apps away from the home screen and into folders to make the icons smaller. You can substitute a helpful app in their location.

- Mute conversations in messaging apps - instead, pre-commit to a time to check and answer your messages (see Building Block 4 - Pre-commitment).

- When downloading a new app, turn off notifications.

- Turn off features such as autoplay.

★ Moving the location of an app on your phone will insert a hurdle, making the associated digital habit more under conscious control (due to having to remember where you moved the app) and, as a result, more fragile. But this is only temporary and it is not long before a new habit for the new app location forms. It is therefore important to deploy this technique along with others described in this book to create healthy digital habits.

Building Block 6: Add Consequences

Reduce your phone's rewarding nature by increasing workload.

- Part of the reason phone checks become our go-to actions is because they are typically brief, effortless and have no added consequences. As such, they are often used to provide a quick escape from difficult tasks.

- Both the Five-minute Rule (Building Block 1) and Inserting Hurdles (Building Block 3) make your phone less rewarding because your brain will apply a degree of delay discounting.

- You can amp this up by attaching a laborious task to certain phone checks – this makes them even less appealing and turns them into a chore. Your brain will realize that brief checks at undesired times create more hassle, making it less inviting and decreasing the frequency of the habit.

Some examples:

HABIT	LABORIOUS TASK
Checking email	Respond to every email in your inbox (if you'd prefer not to appear as though you're responding instantly, you can schedule delayed responses).
	If your inbox is overflowing, decide on a specific number of emails to tackle each time.
Checking the news	Choose a certain number of news articles to read in their entirety despite your interest level.
Social media	Set a number of posts to read/comment on each time you log in.
	Switch your feed from an algorithm-based display to a reverse chronological order.

Building Block 7: Regulate Rewards

Maximize the anticipatory power of dopamine to regulate rewards.

- Start prioritizing quality over quantity when it comes to rewards. Mindfully look forward to fewer, but more significant rewards. Eagerly awaited and earned rewards, fuelled by anticipatory dopamine, bring more satisfaction compared to immediate rewards that quickly lose their appeal due to overindulgence.

- Reframe any frustration you feel when not checking your phone into anticipation. Actively tell yourself that this is something to look forward to.

- Rather than subjecting yourself to numerous, disruptive short checks, use Pre-commitment (Building Block 4) to

---

Strategy

Harness the Power of Anticipation

Maximize anticipation by incorporating this daily anticipatory routine:

- Start every morning by thinking of three things that you are looking forward to, for example, a morning cup of coffee, meeting a friend, engaging in a hobby.

- Feel free to include both digital and non-digital activities - perhaps you're excited about a new episode of your favourite series releasing today, or playing a game.

- Applying this routine not only boosts your mood, but also enhances your enjoyment of these activities when they arrive. It encourages you to be more deliberate with your time and strengthens your ability to savour the anticipation of rewards.

schedule a time where you will engage in your favourite digital activity. Reward planning prevents checks from becoming distractions and trains your brain to eagerly await substantial, delayed rewards, rather than continually seeking minor, trivial ones.

- This approach, particularly when paired with Leveraging Location (Building Block 5) to associate rewards, offers a much-needed reprieve for your executive. Instead of draining your willpower with constant battles against disruptive phone checks, you're providing restorative rewarding breaks in your pre-determined times and 'digital spaces'.

Strategy

Moderate Digital Rewards with the 80/20 Rule

Consider using this rule, also known as the Pareto Principle, originating from economics, which suggests that 80 per cent of the results come from 20 per cent of the effort invested.

- In the context of phone usage, this implies that the majority of value, enjoyment, and benefits you derive from an app probably occur within the first fifth of your interactions. The remaining time likely involves redundant scrolling, offering only minimal additional reward.

- Conversely, for activities like exercising, journalling, or meditating, the greatest benefits can be reaped by making the transition from doing 'nothing' to doing 'something' - even if it's a small amount.

- Embracing this mindset can be a powerful antidote to perfectionism, aiding you in initiating new habits and fostering a balanced approach to your digital interactions.

Step 2: Encoding New Habits

Building Block 8: Really Small Habits

Start a 'really small habit' that aligns with your goals.

- Early on in this book, I asked you to consider what you wish you had the time and energy to do if you could curtail your phone checking – now is your chance to do it.

- No matter which habit you have decided to cultivate, find the really small action equivalent. Remember, this really small action is not your ultimate goal, it is your starting point.

- Ensure that your really small habit is simple enough to be executed in low power mode as it's in these moments we lean most on our habits.

- Remember that your really small habit will be amplified through the 'domino effect'. Sometimes you may perform the habit once – pushing over just one domino – and, sometimes, you will keep going, and there will be a cascade of falling dominoes. No matter the outcome, you have activated your neural networks – you are building a habit.

Apply lessons from technology:

- No action is too small: never worry that your really small habit is too trivial. This way of thinking sets the bar so high that it puts your brain off starting – lower the bar instead.

- Apply constant reminders: these could be based on a time of day, physical location, your internal state or linked to an established part of your routine. A really small action means that it can be inserted anywhere and at any time, so you can form a large number of reminders around it.

- Repetition strengthens the habit: making an action really small makes it more likely to be repeated frequently.

Some examples:

ULTIMATE GOAL	REALLY SMALL HABIT
Read more	Read a single page
Improve flexibility	Do one stretch
Go running	Put on running shoes
Learn a language	Learn a single word
Write a book	Write a sentence
Any large goal	Start for five minutes

REALLY SMALL HABITS

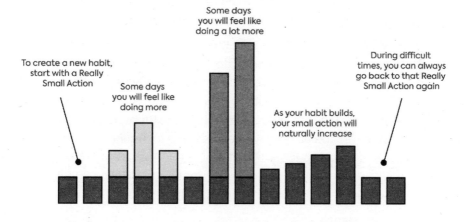

How Really Small Actions Build Habits: Each small action serves as a starting point, laying the foundation for forming long-lasting habits.

Strategy

Flip the Five-minute Rule

Help cement your Really Small Habit to achieve your goals.

- Frame your Really Small Habit as something achievable within a five-minute window to alleviate the pressure and mental resistance associated with starting. Once you've broken past the initial inertia, you might find yourself willing and able to continue beyond the initial five minutes.

- Leverage the same principle as the Five-minute Rule to focus on initiating positive behaviours instead of delaying undesired ones, and to help you achieve an intimidating task by making it more manageable and paving the way for steady, incremental progress.

- For example, if your aim is to exercise more but you struggle with the motivation to begin a full workout, try committing to just five minutes of exercise. It's a small, manageable commitment that can readily slip into your daily routine. These five minutes are a non-negotiable but the decision whether to carry on afterwards is up to you.

- Whether you continue past the five minutes or not does not matter. The key is consistency, not intensity. Every time you implement this rule you activate your neural networks, embedding a habit that will grow bigger over time. Five minutes may not seem like much but, as these small victories add up, they will set the stage for lasting change.

- As you get more proficient with the Five-minute Rule, you are welcome to add ten or fifteen minutes as your target. But, if life gets tough, then reduce it again to maintain consistency.

★ Keep in mind the 80/20 rule – the greatest gains are going from nothing to something.

Advantages of Really Small Habits

- Require little to no motivation, so they will not deplete your willpower.

- There is little to no resistance to doing them which avoids procrastination.

- You can continue to do them in low power mode.

- They can provide a direct replacement for phone checks.

- They activate your neural networks in a similar way to large actions.

- You are more likely to be consistent hence achieve the repetition required to create a habit.

Building Block 9: Substitute Problematic Habits

Give your autopilot an alternative option: replace an old habit with a new one.

- Substituting habits equips your autopilot with an alternative path to take, meaning that it does not enact the problematic digital habit.

- For a transformation to be effective, you need to replace your existing phone habit with a new, minimal-effort action that satisfies a similar need.

- For instance, if your phone habit serves as a mental reprieve during cognitive exhaustion, your replacement activity should also offer a restorative break. Attempting to swap it for a 'productive' activity will be ineffective.

- Ideally, choose substitute habits that incorporate a natural end signal. This way, you avoid the need to exert mental effort in order to force yourself to disengage from the activity when your break is over.

Some examples:

- Instead of scrolling mindlessly through social media, message a friend you'd like to see, read a few pages of a book, water some plants, play a short word game on your phone, plan what you'll have for dinner.

- If you're prone to browsing online shops, tidy up one drawer of your clothes.

- Rather than reaching for your phone first thing in the morning, get up and open your curtains.

Really Small Action x Hurdle

Advanced Really Small Habits

Insert your Really Small Action before certain phone checks as a hurdle. You are not allowed to use your phone until you have completed your Really Small Action.

Some examples:

- If you want to be tidier, clear away ten items before you play a game.

- If you want to read more, make reading a couple of pages a prerequisite to browsing social media.

- If you want to gain upper body strength, do some push-ups before you check the news.

- Use the Five-minute Rule to make a start on a project you've been putting off before watching a new episode of a TV show.

Building Block 10: Temptation Bundling

Increase the rewarding nature of your Really Small Habits.

- Whereas many of our technological activities provide immediate rewards, the benefits gained from non-technological activities such as exercising, studying, doing the housework etc. are often delayed and therefore undergo delay discounting by our brain.

- Temptation Bundling links an activity that provides long-term benefits with instant gratification to foster positive associations which speeds up habit formation.

Try:

- Listening to a podcast to reduce the tediousness of housework.

- Watching your favourite TV show on the treadmill.

- Scheduling your workout right before you have a treat – rewards don't need to be simultaneous to be linked. When our brain encounters rewards, it looks backwards.

- You do not even need to introduce new rewards. Think of times that you'd reward yourself anyway (by indulging in your favourite food, drink, snacks, TV show) and time your supportive habits just before then.

- For every minute you spend doing the habit you want to cultivate, you can allow a minute of access to a particular phone app – rewards can be symbolic too.

★ Reminding yourself that each really small action is a step towards habit formation and is rewiring your brain means you are less likely to engage with your inner critic which can sabotage habit formation and prioritize internal rewards instead.

Struggling with setbacks?

Just this once

- When you're demotivated, your brain, craving a reward, will try to trick you into forgoing what you've practised 'just this one time'. Before you know it, 'just this once' actually becomes a frequent occurrence.

- This is particularly common when you set reminders for social media – you start off by dismissing the pop-up what you think will be just once but end up building a habit to dismiss that reminder every time. It can also happen with the Five-minute Rule or moving to your designated digital space.

- When you're tempted to do something 'just this once', visualize yourself doing the same thing in that exact same situation every time for an entire year.

- Use the Five-minute Rule to assess how you feel about doing this. If you would not be happy doing this every day for an entire year, then don't do it now.

Problem-solve setbacks

- Take the time to think about why you might be struggling with setbacks – they can gift us with valuable information by analysing what happened in a non-judgemental way. Was our willpower low? Were we mentally fatigued, overworked or tired? Or did we simply forget?

- Look for common themes and consider implementing different strategies.

- Use Pre-commitment (Building Block 4) to set an 'if-then' implementation, if the setback was to arise again.

- Remind yourself of the science – researchers find that missing a developing habit for a day or two may temporarily weaken habit strength but this is compensated for a few days later as long as participants get back on track. Over the long term, there are no lasting changes on habit formation.[57]

FORMING HABITS

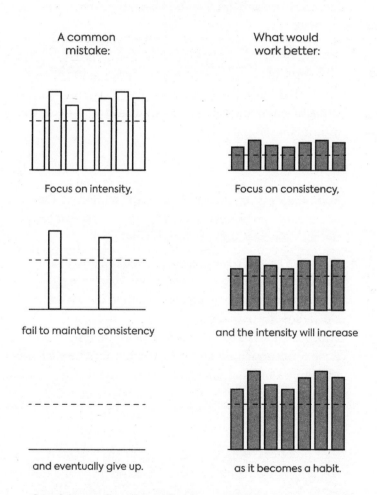

A common mistake:

What would work better:

Focus on intensity,

Focus on consistency,

fail to maintain consistency

and the intensity will increase

and eventually give up.

as it becomes a habit.

Consistency Over Intensity: As repetition is a key part of the Habit Puzzle, prioritising consistency over intensity is a crucial approach to successfully creating new habits.

Track Your Habits

Strategy

Monitor your progress to deal with decreased motivation. We can see our investment in technology - the number of likes or posts, the highest level in a game, the number of followers we rack up. When creating new habits, seeing this progress can be very motivating, so create a metric for your new habits.

- Try 'habit tracking' - noting down every time you apply a new technique from this book. Each one is the equivalent of a small victory.

- Visualizing the repetition you are achieving helps to retain motivation.

Bonus tip:

- Habit tracking can fuel an all-or-nothing mindset. Building up a streak can be very motivating but breaking it can be demotivating enough that people end up giving up.

- If you can relate to this, track your habits through a colouring page: colouring a small part each time you apply a technique, or each day, means you are able to slowly see the page becoming filled with colour. More importantly, missing a day or two does not lose your already accumulated progress which is still there for you to see.

For further practical tools to facilitate your digital habit journey, visit www.drfayebegeti.com, where I'll keep you updated with resources to keep you motivated on your journey. Remember, each small step is a move in the right direction. Hopefully, you have already started implementing some of the techniques from this section. If you haven't, please take reading this as a cue to start now because leaving something until later is to risk not doing it at all. Embrace the process and keep going! In the next section, we will explore in greater depth how technology affects specific aspects of our lives.

PART III

Unlock Your Potential

9

Focus

Our brain has evolved to be naturally distractible, an important safety feature that means we can quickly redirect our attention to immediate surrounding dangers. To be completely absorbed by looking at a leaf would be no good if it left us vulnerable to being preyed on by a wild animal. As our society has gradually become safer – with fewer threats from wild animals at least – the ability to stay focused on a certain task, or on our work, has emerged as a prominent advantage and a trait to be admired. If, however, you were to find yourself in an unsafe situation again, your brain would switch on to high alert, acutely sensitive to noise and movement, readying itself to respond to potential danger. In addition to this important safety feature, rewards are also attention-captivating. For example, you'd want your attention to be diverted from the tasteless leaf towards some juicy and colourful berries. While our immediate environments are typically safe and food is readily available, it is these lingering aspects of our nature that make us prone to distraction and that technology taps into. Our phone provides an online world full of potential perils and rewards that our attention is instinctively drawn to, enabling us to build a range of technological behaviours that we find problematic, particularly when trying to focus or concentrate on a specific task.

A certain degree of our executive function, and hence our focus, is determined by genetics, and so some people will be naturally better at focusing. But although our genes govern our starting point, this does not mean that focus cannot be improved upon. It is a bit like training for a race. Some people have a natural inclination for running but, even if you don't possess this, you can still improve your speed and endurance with the right training and techniques. If you feel you struggle with maintaining your concentration then this can be improved, but you will need to be even more mindful about implementing the right strategies to support your aims. This chapter will teach you how focus works and provide you with the techniques to use technology to your advantage without hindering your concentration.

Procrastination

Getting down to work can be a challenge due to entrenched digital habits, which often make it difficult to get started and increase the temptation to procrastinate. Procrastination is postponing an action despite knowing the negative consequences, and the digital world is a particularly enticing forum for procrastination. Checking devices or apps is one of the most common ways people avoid difficult work in favour of easier digital rewards. Once again, this is a result of a conflict between our executive and our autopilot brain and is most evident when we are in low power mode, as our autopilot is calling the shots and is more sensitive to quick wins. Being surrounded by an abundance of activities that are easier to perform in the moment increases the likelihood of, say, cycling through our frequented apps over starting a valuable task that will have long-term benefits.

Most people are unaware that getting into a focused state requires time and preparation. Athletes go through a predetermined set of warm-up motions both as a physical way to activate their muscles and to get in the zone mentally. For example, prior to each serve, a tennis player will carefully select a ball, walk up to the line, bounce the ball a set number of times and look to a specific position in their opponent's box, before tossing up the ball into the air. In the middle of a match, these routine actions are less about preparing their body than about preparing their mind. In a similar manner, when we are working, we cannot instantly get in the zone. Studies show that people do not start work with an immediate feeling of focus but that this builds over time. For those starting work at 9 am, focus has been shown to peak at 11 in the morning. People will generally then break for lunch and, after an initial sluggish period, focus will peak again mid-afternoon from 2 to 3 pm.[62]

Of course, this pattern is based on averages and each of us will have our own unique rhythm. Again, individual genetics mean that some people may find that their executive brain gets into a focused state earlier in the morning, while others will need a longer warm-up period. What matters here is understanding our own individual pace and managing our expectations without judging ourselves too harshly. Common advice about productivity often proclaims starting with our most difficult tasks, but this disregards the time that it takes to transition into a focused state. Many of us therefore feel that we should be at our best immediately after sitting down at our desk – no warm-up routine required. Under pressure to force this intensity of concentration before our brain has warmed up, we may struggle to get started. We reach for our devices. We procrastinate.

Email-check, news-check, social media-check. These really small actions provide short-term rewards. Over time, and with repetition, this digital warm-up routine becomes encoded in our autopilot brain and bound to the reminders of our work location. Studies show that a substantial amount of procrastination is out of habit,[63] and those habits mean that it doesn't feel right to start some of our more difficult work before making sure that we are fully up to date with our most frequented apps. We become just like the tennis player preparing to serve, except our preparation is our own personal digital cycle.

This digital warm-up routine can be counterproductive, creating unhelpful digital actions in what should be our focused location, and which start to invade other aspects of our work.

Unlike a finite warm-up, the absence of clear stopping cues in the digital world renders our time there unpredictable, enabling us to elongate or repeat our digital distraction cycle as a means of procrastination. The volatility of the virtual world can impact our focus and productivity in significant ways, causing us to start our workdays on other people's terms rather than our own. For instance, when we open our inbox, we are greeted by a multitude of messages from colleagues, clients, or even promotional offers, all competing for our immediate attention. This bombardment sets an agenda for us that we didn't choose; it's dictated by the demands and needs of others. Similarly, when we browse social media as part of our digital warm-up, we are influenced by the posts we see, whether they're news updates, personal statuses, or opinion pieces. These messages can stir up emotions, which require executive power to emotionally regulate, as well as irrelevant thoughts that may side-track our focus. This influx of unfiltered information can set the tone for the rest of our day, affecting our mood, our thoughts, and even the course of our work. Instead of starting our day with our own defined goals and intentions, we end up reacting to the virtual world's constant flow of information and demands, which can be chaotic and distracting. Moreover, stopping this cycle takes a significant amount of willpower, so we find ourselves depleting precious mental energy before we have even started.

You'll remember that delay discounting in the brain means distant rewards get artificially reduced. In the morning, with the whole day stretching ahead of us, there is less incentive to make sure a task is completed by the end of the day, let alone to meet a deadline in the distant future. Our autopilot instead steers us towards short-term rewards and is in direct conflict with the long-term plans of the executive. We scroll on our phones even though previous experience may have shown us that we will feel worse about this later. Then, as a deadline looms, our short-term thinking autopilot starts to align with the executive. Having a short-term reward to look forward to possibly explains why, for many, there seems to be a burst of focus prior to an anticipated lunch break, or mid-afternoon prior to going home, both of which provide artificial deadlines, and have been shown in studies to focus our mind and reduce procrastination.

Digital procrastination isn't solely related to initiating focused work; it can creep in anytime we confront something challenging. Many aspects of our lives necessitate a degree of hard work, often inducing a level of discomfort our brain instinctively wants to avoid. Activities like studying, exercising, and cleaning come with inherent difficulties. Similarly, emotional tasks such as having a difficult conversation, making a challenging phone call, or even tackling a difficult project that we fear might end in failure can be daunting. These activities have a common thread: they demand effort and/or emotion regulation from our executive, making them prime targets for procrastination. There is a degree of resistance to starting and, when this resistance gets too high, we seek to avoid. Again, overcoming this resistance requires executive power and when our executive is already fatigued, we are even more likely to defer. We often procrastinate when it comes to starting our workout, tidying up, or even going to sleep, a phenomenon we'll delve into further in Chapter 10. Our autopilot brain seeks short-term rewards, and in doing so, as the Habit Puzzle is completed, we encode problematic habits. These embedded scripts are then often enacted automatically, making it progressively more challenging to initiate these tasks in the future. Phones, in contrast, provide a very low barrier to access information and entertainment, with no associated hard work attached to them. There is little to no resistance to picking up our phones and this is why we have developed behaviours of picking them up more than any other activity.

When faced with a difficult task, one that we want to procrastinate, our brain fools us into thinking that the energy needed to complete it will be expended equally throughout. Therefore, we assume that the amount of energy required is proportional to how long or how difficult a task is. This is not the case. In fact, disproportionately more energy is required to start and, once you have begun and have some momentum, significantly less energy is required to keep going. The challenge, therefore, is getting started, and is why most people procrastinate. Using the Five-minute Rule (Building Block 1) is an effective method of overcoming this habit. You can commit to work on a task for just five minutes and give yourself permission to procrastinate afterwards. It is not uncommon to find that, once we overcome the initial hurdle of getting going, we are more likely to continue the task than to stop. Even if you do stop afterwards, you have started the act of substituting your procrastination habit (Building Block 9), giving you a chance to activate your neural networks and start rewiring them. You can also incorporate Plan B (Building Block 2) meaning that you start with a less mentally taxing task to give your brain time to warm up before tackling more the more challenging Plan A. This is a tool that I have used myself. It takes me considerable time to get into the flow of writing so, when I needed to work on this book, I started by passive reading and editing what I had written the previous day. Rest assured, implementing Plan B here is not taking the easy way out; rather, it is a training method for your executive. Think of it as doing a light warm-up before lifting heavy weights at the gym. Remember, your Plan B should be an active, conscious choice and an activity that aligns with your overall goals, rather than one that promotes procrastination. While you make getting started easier for yourself, you can simultaneously make digital procrastination more challenging. By Inserting Hurdles (Building Block 3) to increase friction in your digital habits and Adding Consequences (Building Block 6) you can further reduce its appeal.

Since procrastination arises as a result of a conflict between the autopilot and the executive, it is helpful if distant deadlines are converted to something tangible in the present. Unfortunately, many people make the mistake of setting deadlines that are too far away to engage the autopilot brain. For instance, breaking down a year-long project into three-month intervals is not compelling enough for the autopilot brain to take notice. To truly get the autopilot brain on board, it's best to set artificial, short-term deadlines which are no more than one or two hours away. If you are really struggling with your focus, those deadlines should be even closer, and you can gradually increase them with practice – again, just like you would increase weights at the gym. Once completed, rewarding yourself at the end of each deadline is key as the autopilot brain is drawn to short-term incentives. This is why the participants in the studies above experienced a surge of focused attention in the hour just before lunchtime or at the end of their working day, when they knew they could take a break or go home. You too can leverage this effect by scheduling a small reward (e.g. taking a break or having a snack) after completing a defined period of focused work. The reward can also be digital, such as browsing some online shops or listening to a podcast. The aim is to build a habit where you mindfully engage in earned rewards, utilizing the motivational component of dopamine. Practising anticipation is important as the act of looking forwards to rewards magnifies their effect and helps regulate them (Building Block 7). Ensure that you pause and congratulate yourself when meeting goals, something that is often forgotten, no matter how small they are. Rewarding your periods of focused work, while concurrently making it more difficult for yourself to procrastinate, will help tip the scales in favour of both more intentional work, and more intentional phone use.

The impact of interruptions

When asked about the reasons that phones impede our focus, the first thing that comes to most people's minds are interruptions – the almost constant notifications that draw your attention away from a task, such as WhatsApp messages, news alerts and new emails pinging into your inbox, and which you might as well quickly read before you get back to work. An essential requirement for our brain to enter a focused state is a sufficient period free of interruptions. As discussed above, it takes time for our brain to enter the deeper stages of focus so, if we are regularly interrupted, this fragments our attention and disturbs our train of thought. We may therefore end up doing most of our work on a superficial level and studies show that the attentional lapse after receiving a message increases the probability of us making a mistake.[64] Interruptions can also be more time-consuming than we expect with research showing that, after a significant disruption, it can take us an incredible twenty-three minutes and fifteen seconds to return to our original task.[65]

Interruptions can be subdivided into two categories. The first, external interruptions, are disturbances from people or systems outside yourself, like phone calls, messages or notifications. Most of us will be well aware of their potential to distract us from valuable work and we may even loathe them. Then there are internal interruptions, when we disturb ourselves in the middle of a task before a natural break point has been reached, and which so often go unnoticed. In Chapter 5, we saw that only 11 per cent of the time people check their phone is due to an external interruption such as a notification.[19] The overwhelming majority of the time, people will reach for their phone completely unprompted – this is a self-interruption. When trying to do focused work, studies show that self-interruptions occur at nearly the same frequency as external interruptions,[65] except that we are less aware of them – a sign that we are acting on autopilot. Self-interruptions can happen as often as every few minutes for some. Given that we cannot avoid ourselves in the same way that we can create space from other potential external distractions, the constant presence of these self-interruptions means they have the potential to be much more disruptive.

*

You might recall from Chapter 5 that we discussed 'reminders' which fall into both external and internal categories. At first glance, they might seem synonymous with interruptions. However, they serve different functions. Reminders serve as cues to the autopilot brain to initiate a particular habit. On the other hand, interruptions, both external and internal, halt an ongoing activity, disrupting our focus and flow. Sometimes there's a bit of confusion between these terms because reminders can cause us to interrupt ourselves. For example, seeing our phone could act as a reminder of the habit of checking social media, and we might stop what we're doing to scroll through our feed. But it's important to understand the difference: a reminder is like a green light saying 'Start this action', while an interruption is a red light saying 'Stop what you're currently doing'. When a reminder leads to us stopping a task, that's when it has resulted in an interruption.

Self-interruptions are more likely to occur when our executive is fatigued. In these instances, we may find ourselves in low power mode but, instead of taking a break, we push on through. Remember, in this state the executive brain cedes control to the autopilot, which is drawn to quick rewards. External interruptions can also contribute to this phenomenon. Studies show that individuals who experience interruptions outside of their control are more likely to self-interrupt later on[66] – this is because such external interruptions break down a person's attentional stamina and fatigue the executive, so making them more prone to self-interrupt. Over time, self-interruptions, like procrastination, can become a habit and we end up checking our phone or engaging in other distracting activities without realizing.

Having stored self-interrupting habits, our autopilot brain is prompted to execute them through reminders in our surroundings or our internal state, even when we are not in low power mode. To help break this cycle, identify the specific actions and activities that you self-interrupt with and try Inserting Hurdles (Building Block 3) which will halt autopilot mode and activate your executive brain. Then, to test whether an urge to self-interrupt is an automatic response or a genuine need for a break, try implementing the Advanced Five-minute Rule: Surf the Urge (Building Block 1). If the urge fades within five minutes, it was likely an automatic response, and you can refocus on your work. Implementing this rule over time builds your executive stamina and makes you less likely to respond to immediate impulses. However, if the urge to self-interrupt persists, this may be a sign that your executive needs a brief disconnection. A quick task shift, when used wisely, can give our fatigued executive the break it needs.

For many people, self-interruptions are a sign that their brain is cognitively fatigued. When our exhausted executive is trying to take a break, we often navigate towards less challenging tasks involving our autopilot. However, in our effort to be productive, we are not necessarily providing ourselves with a particularly restful pause. A common example is a self-interruption to check email – we open our inbox to see what is in there but, too fatigued to reply, we leave any new messages to fester, putting further strain on our executive as multiple unfinished tasks loom in the background. This is inefficient and we return to our work feeling even more fatigued after this 'faux break' hence being more likely to self-interrupt. To combat this, really differentiate what is work and what is a break by Adding Consequences (Building Block 5) such as having to reply to all the emails you receive as soon as you read them.

For a break to be restorative, it needs to adequately give our fatigued neural networks a chance to rest. As discussed in Chapter 3, this will be individual and, in part, depend on what we were doing previously, so you need to keep in mind whether your chosen activity is energizing or depleting for your executive brain. The renowned writer Maya Angelou would often have a pack of cards so that, when she couldn't write, she would have something else to occupy her 'little mind'.[67] This phrase was passed on from her grandmother, and is a very astute way of describing the autopilot brain. Strategic breaks, where we occupy our autopilot with an instantly rewarding and less challenging activity, are able to provide temporary rest for our executive (like pausing between sets in the gym). We can then go back to our work feeling refreshed and with new ideas. You can still choose a digital activity for your break, but you need to ensure that it is helping you recharge. For example, browsing the news or social media may be a welcome break for some, but for others, processing argumentative comments or negative news stories, which require a significant emotion regulation component, may drain their mental energy. In such cases, substituting this activity with a different digital or analogue alternative will provide a more efficient break.

A key rule to have in mind is that intentional breaks are restorative whereas distractions are fatiguing. Numerous self-interruptions do not add up to a satisfying break and are instead a source of frustration. This is partly because of the guilt associated with getting distracted and the willpower that we further expend as we try, and repeatedly fail, to stop scrolling and get back to work. Self-interruptions can be so frequent that the time spent pursuing momentary distractions could have instead been used for a meaningful, replenishing break. This is similar to when someone is hungry but ends up snacking, and in doing so eats the same amount as they would have when eating a proper meal. By opting not to take breaks, we wait until we succumb to distractions, which end up lasting long enough that a restorative break would have been more time efficient.

As we saw in Chapter 5, location plays a key part in how our autopilot executes our habits, but if we spend significant amounts of time self-interrupting at our desk, it may no longer represent a work area. It's no surprise that our autopilot brain is getting confused. Maya Angelou instinctively recognized this and would go to a hotel room to write, a place which only signified work for her. When her executive needed a break, she would switch from the desk and sit on the hotel room bed to play a game of solitaire, which acted as a reward for her autopilot brain. This is sound advice that I recommend. Define a place of focus and use a different location, even if it is within the same room, for intentional breaks (Building Block 4) to help control the reminders in your environment.

We need to manage the unrealistic expectations we put on ourselves of working for long focused sessions non-stop, and which are a key influence on why those self-interrupting habits develop in the first place. Angelou would start her writing process at 6:30 am, and would call it a day at 1 pm, setting her expectations ahead of time, actively stopping her work. We often work continually, until the point of exhaustion, honing a lot of self-interrupting habits in the process. If you find that breaks are no longer restorative and you are self-interrupting with increasing frequency, then it may be the case that you have now found yourself in low power mode and should call it a day. The idea of doing this might feel counterintuitive to our beliefs about productivity and time management, but pushing through low power mode and continuously self-interrupting means we are wasting mental energy to re-orient our attention back to our original task. This is much like the fuel inefficiency of completing a car journey while starting and stopping multiple times. Having to constantly stop and redirect our attention prevents us entering the deepest stages of focus. Working with a fragmented attention span results in superficial and lower quality work, despite the long hours put into it. This can impact different aspects of our lives. In an academic setting, it could mean struggling to retain a concept due to shallow understanding. In another working context, it might undermine your capacity to assess a problem and prevent you from coming up with innovative solutions. Ultimately, it's about redefining our perspective on productivity. It's not about long hours filled with continual disruptions, but fewer, more focused hours that allow us to dig deeper into our tasks. This not only ensures higher quality output but fosters a healthier and more sustainable approach to work.

Multitasking

In an increasingly busy world, where we are battling constant interruptions and the urge to procrastinate, it is not uncommon for us to try and do multiple tasks at the same time. This is often encouraged by the notion that it will increase our productivity and somehow help us make up for any lost time. However, the neuroscience examining multitasking shows this to be a myth. When we multitask, we think that we are doing two things simultaneously, but that's not what happens. Our brain is instead quickly flicking attention from one task to the other very rapidly – something called 'attention switching'. Rather than being productive, this actually has the opposite effect. It takes greater effort for the executive brain to constantly switch attention than to concentrate on one item at a time and complete tasks sequentially. Therefore, multitasking is actually a mentally draining tactic which reduces our efficiency.

Most people think
they can do two tasks
simultaneously

However, instead of
multitasking we are usually
attention switching

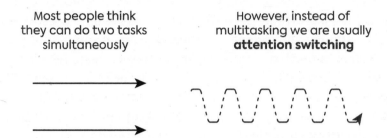

Multitasking vs Attention Switching: For the vast majority of people, multitasking – trying to do two things at once – in fact involves flicking attention between two different tasks, which reduces efficiency in the process.

To experience the detrimental effects of multitasking firsthand, try this simple task: write 'I am a great multitasker' and the numbers 1–20 on two different pieces of paper. First, try doing so simultaneously, switching rapidly between each letter and number. Then, write the sentence and numbers again, but this time do the tasks sequentially, completing one before starting the other. Compare the time it took and the difficulty level of each approach. I would wager that the sequential task is both quicker and easier than rapidly alternating between the two. This exercise demonstrates the inefficiency and difficulty inherent in multitasking. It is important to note that multitasking has less to do with which muscles you are using than the ultimate aim of the task you are seeking to accomplish. For example, playing the piano may require many muscle movements to be done simultaneously but it would not be classed as multitasking because it seeks to achieve the same purpose – producing music. In contrast, trying to multitask two activities with different aims, such as checking unrelated emails during a seminar, results in the ineffective process of attention-switching.

The integration of technology into our lives has led to the rise of a more recent phenomenon: media multitasking. This is when we engage with two or more types of media at the same time, for example, watching a film while scrolling on our phone, watching a YouTube video while trying to simultaneously read the comments or instant messaging friends during a webinar. Wanting to understand the impact of this on the brain, a highly influential study in 2009 pitted the brains of heavy media multitaskers against those who did not media multitask.[68] The test was as follows: two red rectangles would appear on a screen for less than a second. There would be a small pause, then the two red rectangles would appear again. What the participants had to do was to indicate whether those two red rectangles had changed orientation. This task required laser-sharp focus and participants had to keep their attention on the screen at all times as letting their mind wander meant they would easily make a mistake. But then, the researchers made the task even more challenging and added some blue rectangles into the mix. The blue rectangles were completely irrelevant, and all the participants had to do was ignore them, but for one group, this was easier said than done. When only the red rectangles were present on screen, both groups were able to focus and complete the task. However, as more blue rectangles were introduced, the heavy media multitaskers' performance dropped until it was significantly worse than those who did not media multitask. It wasn't their focus that was the problem; it was their ability to ignore distractions.

Broadly speaking, people tend to fall into two camps: those who feel that they cannot multitask and those who are proud of their ability to do so. Scientific studies show that our perception of our ability to multitask is badly inflated. In one study, the majority of the participants recruited estimated their multitasking ability to be above average, something that isn't mathematically possible. The reason for this became clear during formal testing where scientists found a disconnect between perceived and actual multitasking ability, concluding that 'participants' perceptions of their multi-tasking ability were poorly grounded in reality'.[69] Turns out thinking that you are good at multitasking does not mean that you perform better – in fact, you may do worse.

The reason why is logical; doing complex tasks depends on the executive brain and it therefore follows that you need high levels of executive function – the ability to focus, concentrate, and block out distractions – to be able to do multiple tasks together. Ironically, those people who would naturally be better at multitasking are actually less likely to do it, as they tend to concentrate on a single task at once and are distracted less. Those more likely to multitask in real life had lower executive control and scored lower on the multitasking test in the lab. Multitasking has also been shown to significantly correlate with impulsivity.[69] Rather than intentionally multitasking, chronic multitaskers may do so because they have difficulty focusing their attention on a single task.

The above studies raise a key question – does media multitasking increase distractibility, or do inherently more distractible individuals (possibly due to genetic factors) gravitate towards media multitasking? While science cannot yet provide a definitive answer, we can nevertheless take action to mitigate potential issues. For example, creating an environment with fewer distractions can be beneficial regardless of whether distractibility is a cause or an effect of media multitasking. These early studies should also make us reflect on our own behaviours by considering the following questions: Is media multitasking a deliberate choice, or a symptom of your inability to ignore the technology-based distractions? If you scroll on your phone while watching a film, are you doing so intentionally, or has it become an automatic habit? If you check and respond to emails during a lecture, are you genuinely capable of handling both of those tasks concurrently, or are you simply unable to wait until the lecture is finished?

This does not mean that you should never multitask, but you should be aware of its pitfalls. As with much of the approach of this book, it's about finding the right balance. The golden rule to effective multitasking is that one of the tasks must be simple enough to be performed automatically. This means that the executive brain can delegate the task and focus on something else without being drained by constant attention switching. The type of tasks this includes will vary from person to person. For example, the reason most people can walk and talk simultaneously is because our walking becomes automatic through pattern generators in our spinal cord. Walking will require full concentration for a young child who has not developed those connections just yet – my youngest daughter who, at the time of writing, was just learning to walk, would easily fall over if she got distracted. Similarly, walking a familiar route can be navigated by our autopilot while we maintain a conversation using our executive brain. However, when we come across an unexpected diversion, we usually find that we need to pause our conversation because we require executive input to adapt to an unfamiliar route.

Processing, understanding, and retaining information will always involve the executive brain and cannot be done on autopilot. This is why media multitasking has limitations compared to other forms of multitasking where jobs can be delegated to other brain regions. Trying to listen to someone speak while checking your emails is both ineffective and fatiguing, as is trying to study while watching TV, trying to immerse yourself in a film while scrolling through a social media feed or trying to watch a video while reading the comments. I personally most commonly multitask by listening to podcasts or audiobooks while doing household chores – things I've done many times before, like unloading the dishwasher, which can be done on autopilot. However, if I start organizing my cupboard, a task which requires more complex decision-making using my executive, I am not able to multitask effectively, and the end result is that I stop paying attention to what I was listening to.

Being able to technically do certain things simultaneously does not mean that performance is equivalent to monotasking. I will process the information from the audiobook at a much more superficial level while doing housework than if it had my undivided attention. Multitasking likely also reduces my cleaning speed and increases the probability of making an error – putting things away in the wrong location, for example. However, assuming that performance is not important, and any error will not be critical, multitasking can still be a useful tool to increase the rewarding nature of an otherwise-dull activity (Building Block 10). But it is helpful to remember sometimes that just because we are able to multitask, doesn't mean we should. It is OK to give our brain some space in an increasingly demanding world. Monotasking can be a form of mindfulness and a time to focus on the present. Alternatively, you can let your thoughts wander. When you do that, something very interesting happens.

The default mode

To figure out how the brain works, subjects are frequently invited to lie down in powerful brain scanners where they undertake numerous mentally-challenging tasks in order to test aspects of their thinking. When an area of the brain becomes more active, its metabolic requirements increase, leading to a rise in blood flow that can be detected through imaging. However, for a long time, researchers overlooked an important aspect of brain activity. Prior to each experiment starting, there would typically be a brief period where participants were lying still, waiting for the task to start. During that time, scientists observed that their subject's brain, rather than being quiet, was actually very active. When the experiment started, those regions of the brain that were firing would fall silent.

Even when we are resting, our brain activity does not switch off. When we allow our mind to relax, different parts of our brain are active. This effect is not limited to a particular brain area but rather a network of brain regions with organized patterns of activity. This network is active at rest but becomes deactivated during a focused state. It is what we experience as 'mind wandering'. In addition to rest, we engage in mind wandering during straightforward, habitual tasks such as getting ready, doing housework, or commuting. During those tasks, we allow our brain to operate on its default setting, a phenomenon that inspired the name of this network of brain regions that become active – the default mode network.[70]

The interconnectedness of the brain regions that form the default mode network means that they can use our stored knowledge and experience and combine them in novel ways to form powerful new ideas. When we are doing tasks that do not require our full concentration, our mind floats between different thoughts and continues to work on previous problems we've encountered in the background. This is why you might find that you get creative ideas or solutions to past issues seemingly out of the blue. It is why when we cannot reach a decision, we may say that we'll 'sleep on it'. We let a problem marinate so that we can work on it in our default mode.

Using our smartphones to be productive or to be entertained when we have a spare moment may be a reasonable thing to do some of the time, but we also need to have space to think. Space to be bored. Space for our mind to wander. Space for our default network to be activated and to form new ideas and new connections. Constantly consuming content in every possible scenario and in every pause in our day means our brain uses resources to process this and silences our default mode network, impeding its background problem-solving ability in the process. There is a reason that so many ideas come to us when we are having a shower, going for a walk, or doing something else mundane.

Life is full of awkward pauses. We may wait for the kettle to boil, a file on our computer to load or the next train to arrive, and, sold the narrative that we must occupy every single minute when we are awake, we use our phones to do just that. This is not an all-or-nothing scenario: using technology to complete a task during some of life's pauses can allow us more time off elsewhere but we must recognize the limitations of filling every single moment in our day. This time when we are not pressured to be productive may end up being our most productive time, a hidden cost in that the more we consume content, the less able we will be to create our own. That small observation while waiting for that train may have been a great idea that was missed.

Misusing technology

Technology provides us with powerful tools. A smartphone is a multi-purpose tool that, when used correctly, can help our executive in many ways. Being able to take a quick note, set a reminder, or look something up instantly means that there is less information that we need to keep in mind. This reduces the strain on our working memory, a key function of the executive brain, and phones can therefore increase the availability of resources that our brain has to focus on the task at hand. However, we are often using our tools in such a way that they burden – rather than help – us. In addition to self-interruptions, constant multitasking and having little time for free thought, the advent of the smartphone enables us to carry our work with us everywhere, to perform tasks on the go and at any moment, blurring the boundaries between work and rest – and the world we live in perpetuates this. It would be far too simplistic to think that the digital habits we have formed around technology are purely individual; there are wider societal factors at play.

The advent of the smartphone, instant messaging and email has meant that communication can be asynchronous. That is, rather than both parties having to stop what they are doing to communicate, we can respond to requests in a time that is convenient for our own focus and schedule. Large parts of society, in particular introverts or those who are neurodiverse, may prefer to communicate via text or email, which also allows the time and space to craft careful, more thoughtful responses. However, being constantly available or 'online', and answering email and messages quickly, has become one of the most visible aspects of productivity. Pressure to maintain these new social norms not only leads to rushed replies but also to forming multiple self-interrupting and multitasking habits that are not serving us well in the long run. In particular, our email-checking habits have turned email interaction from asynchronous into synchronous communication. A study observing a group of employees showed that email is a powerful source of both external interruptions, via notifications, and self-interruptions. Workers, on average, checked their email client more than seventy-four times per day or around eleven times per hour.[71] We frantically try to get on top of our inbox but the quicker we respond, the more follow-up emails are likely to be sent. Unlike traditional mail, which was delivered in batches, email is delivered constantly with a large proportion of people leaving their email client permanently open, only to be distracted by every ping or, alternatively, to self-interrupt to check what's there.

The true impact that this constant availability and blurring of the boundaries between home and work has on our executive brain is something that is widely underestimated. I know the effects of it too well. As a doctor, I often do twenty-four- or forty-eight-hour 'on-calls' where I can be contacted to give neurology advice at any moment, even when I go home. During those days, just the knowledge that I can be interrupted at any point means I cannot completely focus. This division of attentional resources was shown in a study of 520 undergraduates who were better able to solve complex puzzles when their phone was placed in another room than when it was face down on their desk.[72] This is because, even when there are no notifications, our brain will devote a certain proportion of our resources to monitoring our phone for the possibility of a disturbance. This effect, dubbed 'brain drain' by the authors of the study, was shown to be greater in those who have a problematic relationship with technology. This state, equivalent to my medical on-calls, has the effect of reducing our capability to get lost in a task. Parents may well have an inherent understanding of the consequences of being constantly 'on call'. It is hard to give an activity our undivided attention when our children are being looked after in our home compared to when they are in a day-care setting. Our brain still devotes significant resources to monitoring their well-being even when they are sleeping. It turns out our phone is no different and allowing ourselves to be at its beck and call is one of the ways we use it incorrectly.

Does this mean we should abolish, or seriously curtail, access to our devices all together? A 2011 study which experimented with cutting off email for a few employees found that they were able to focus more intently on their work and spent less time switching between tasks. They wore heart monitors, which showed that their heart rate variability had increased, a sign of reduced stress levels and improved cardiovascular health.[73] Superficially, it seemed like email was the problem and cutting it off was the solution. However, investigating further, the findings of the study had a more compelling explanation. Email was not the problem; how people were using it was. For example, one of the participants, a lab scientist, who had reported that email was creating so many interruptions with menial tasks that it was interfering with his ability to run experiments, stated that these tasks had completely ceased. This was not because he was inaccessible. He was, in fact, a short walk down the corridor from his supervisor. But email, used in the wrong way, creates additional unnecessary work. The simplicity of reaching out to someone via email makes it fall into the category of really small actions. This ease can foster a tendency to delegate tasks, often leading to excess communication compared to investing a bit more effort to solve a problem independently. The development of such email-sending habits can artificially increase our workload, leading to stress as well as encoding email-monitoring habits which may impede our focus.

A significant explanation of the findings of this study was that the elimination of email led to a reduction in workload. This suggests that the apparent efficiency brought about by email and various aspects of smartphones may, in fact, be masking an underlying issue of increased workload. Things that would previously take time can now be done very quickly. For example, rather than a trip to my university library to read a scientific paper, I can download it in an instant. For my job, my smartphone provides easy access to whatever I want to look up for my patients, whether it is a drug dose or the latest guidelines for treatment. The problem is that this efficiency does not necessarily counteract the growing demands and complexity of our work. We now have fewer easy routine tasks that can be done on autopilot, such as the walk to the library, or travelling to meetings, which reduce the strain on our executive. While the time saved through the efficiency of technology should provide opportunity for more mental breaks, we instead take fewer. A striking example emerged during the COVID-19 pandemic. The absence of commuting – a time typically used by our brains to decompress and process the day's events – was filled, on average, with an additional 48.5 minutes of work. This extra work time exceeded the duration of a typical commute.[74]

We can now fit tasks into awkward pauses, eat lunch while scrolling, and use our commuting time to check emails, which can all lead to a feeling of being overwhelmed. For most of us, this is not a personal choice but a response to outside pressures. Many people are monochronic – this means that they prefer to work on one task at a time without interruptions – and so these multitasking, time-filling behaviours have been imposed upon them. The solution to our ever-increasing workload and busy lives, often with parental and caring responsibilities, tends to be to find ways to be more productive. Yet the emphasis on constant productivity unfairly blames individuals when they can't keep up. We do not berate people for not being able to run an ultra-marathon but, often, not being able to achieve pre-set levels of constant productivity is seen as some sort of personal failure. The truth is, we all experience executive fatigue, though it manifests at different points for each individual. This variance is shaped by a combination of factors. Genetics play a significant part, but our habits, sleep patterns, and stress levels also contribute – and the truth is that some of these factors are simply beyond our own personal control.

Moreover, our endless quest for productivity means we are developing bad digital habits as a direct effect of being tired, overworked and under-slept. When executive fatigue sets in, we enter low power mode. In this fatigued state, rather than tackling a difficult task on our to-do list, checking our phone provides a relatively easy win. The result is the encoding of problematic habits into our autopilot brain that further conflict with our goals and which we have to fight against. These problematic habits reduce our efficiency, slow us down and add to the overall feeling that we do not have enough time to take breaks. In an attempt to compensate for this decreased efficiency, many of us find ourselves working longer hours, taking on too much and sacrificing precious sleep, and just like inadequate rest can lead to physical injury, insufficient mental rest can eventually lead to burnout. Contrary to our goals, these factors lead to us becoming less focused and less productive.

Completely cutting off our access to our devices – or taking periods of abstinence – is clearly not the answer and nor would it be practical. We therefore need to find a way to use technology wisely and to make our lives easier. This is why I do not condone a 'cold turkey' detox approach to phone use and argue we need to be mindful of our individual habits, work on our personal executive stamina and ensure we use technology in the right way. Constant pressure to be productive is not a recipe for focus but a cause of distractibility due to a fatigued executive. My intention for this book is for it to be as helpful as possible for you but relying solely on these methods and ignoring the wider context of our pressured society can at times feel like applying a temporary bandage without addressing the root cause. Societal change takes time, and transforming ingrained patterns and expectations will require collective efforts, policy change, and a shift in how we prioritize work-life balance. We will need to advocate for healthier work practices and recognize collectively the value of downtime and recharging our executive brain in a world that relies increasingly on technology. However, understanding individually how our brain works and why we form these powerful distractibility habits around our phones – rather than solely blaming the technology itself – is a fundamental step in the right direction.

Focus Practical

Positive actions to improve focus

LEVERAGE LOCATION (BUILDING BLOCK 5)

- To maximize the location properties of habits, define a place of focus and only remain there for work.

- Just having your phone next to you fragments attention, so put it in a different room. When you want to check your phone, move to a different location.

- If you work in a nomadic way, or at a temporary desk that usually serves an alternative purpose, like the kitchen table, then use a meaningful object to indicate that your focus time has begun. Good examples are a framed quote, a picture of your loved ones or an object that has personal meaning; I use a little owl. The type of object is less important than using it consistently as your brain will eventually learn to associate this with focus, encoding it as a significant reminder for your focus habits.

ADJUST YOUR CALENDAR FOR FOCUS

- Constantly shifting between tasks demands cognitive energy and can lead to mental fatigue, making us more prone to procrastination and digital distractions.

- Fine-tune your calendar by blocking out 'focus time' each week. During those times, avoid task-switching wherever possible.

- Split your day into large blocks and keep tasks with a similar theme together to reduce wasting energy from having to stop and change direction. For example, rather than working on a project for an hour a day only to switch to something else, it may be more valuable to set aside two hours for the project on alternate days or even dedicate a longer block of time each week (taking into account executive fatigue constraints).

- To counterbalance the periods of focus, block off time for less taxing tasks like admin. These blocks may be best placed on days that already have a lot of interruptions such as meetings.

- Admin can also be a good Plan B (Building Block 2) for times that you are in low power mode. Having a dedicated admin block creates a space to slot in minor tasks that can be done later and avoid self-interruptions.

BUILD EXECUTIVE STAMINA THROUGH REPETITION

- Developing executive stamina is not about one-off large actions but about consistency, and to build consistency it is best to start small. You need to go on several runs to improve your running fitness and stamina, and the same is true of focus.

- Start a Really Small Habit (Building Block 8): rather than having an amorphous amount of time to do focused work, such as the whole morning, try starting small with fifteen minutes. If you can only complete a little focused work before needing a break, that's fine. This provides a baseline which can be built upon.

- Take adequate breaks: stretching yourself to spend a long time at your desk while not focusing is not useful and risks building contradictory habits. It's like completing an exercise with poor form and will not increase strength.

Tools to tackle digital procrastination

HARNESS THE POWER OF REALLY SMALL HABITS (BUILDING BLOCK 8)

- Use the Five-minute Rule to overcome the initial starting hurdle. Commit to starting your task just for five minutes, then assess how you feel afterwards. Allow yourself permission to procrastinate, if needed, but only after you've put in these initial five minutes.

- Keep in mind: starting a task requires more energy than continuing it. More often than not, you'll find yourself wanting to press on once you've begun.

- Over time, this tiny habit can be amplified through the domino effect, gradually replacing any digital procrastination habits.

ADOPT A PRE-FOCUS WARM-UP

- You may be procrastinating because your brain cannot achieve the intensity you expect immediately – so you end up putting off getting started.

- Create a focus warm-up routine before your deep work sessions: gather everything you need in your workspace, maybe water the plants on your desk, jot down an affirmation, or prepare a hot drink. The latter can feel rewarding, particularly if you save a special blend or brand for these focused sessions.

- If you incorporate digital activities into this routine, make sure they are self-limiting and come with natural stopping points – for example, I often played a game of Wordle to jumpstart my brain before I started writing this book. The beauty of it is, you can't play more than one game per day. However, it's important to steer clear of digital activities that have morphed into troublesome digital habits or those you frequently turn to as part of your digital distraction cycle.

- The exact actions matter less than applying them consistently to help to build Really Small Habits (Building Block 8). With repetition, your brain will associate these actions with the need to prepare for deep focus and use them as reminders to shift gears.

USE PLAN B (BUILDING BLOCK 2)

- If you are struggling to get started, rather than reinforcing procrastination, execute your Plan B (Building Block 2).

- Remember: you do not always need to do your most intense work first if you are not able to. Plan B does not mean you have failed.

SET ARTIFICIAL DEADLINES

- Convert distant deadlines into tangible, near-future goals – within one or two hours – to capture the autopilot's attention and engage it with the executive.

- You can adjust to shorter intervals if focus is a challenge, and gradually increase the time frame.

- Tie these deadlines to rewards to incentivize the autopilot and further synchronize it with the executive.

How it works in practice:

- 'I need to work on this project for the next hour because I need to meet my friend for lunch.'

- 'I will spend the next thirty minutes doing some studying and then I will catch up with my friends' Instagram stories.'

Strategy

The Daily Wind-down Ritual

Cooling down is just as crucial as warming up, yet most of us seldom make an active choice to stop working. Instead, we often wait until we're exhausted and unproductive.

- Try a cool-down routine to signify the end of your working day and establish boundaries.

- Take a couple of moments to non-judgementally reflect on what went well and what you learned. You could leave a note for yourself to read during your next session, such as a positive affirmation or a task that you should prioritize.

- Set up your space - both physical and digital - for the next day. Clear the tabs of any websites you may have opened on your desktop and tidy your desk. When you start work, it is important not to get inadvertently absorbed by the leftovers from the day before.

Managing interruptions

USE HURDLES (BUILDING BLOCK 3) AND CONSEQUENCES (BUILDING BLOCK 6)

- Insert Hurdles to halt autopilot mode and activate your executive, which gives you a chance to consider your goals. Make sure you log out of any news or social media apps that you typically use for self-interruption.

- Introduce Consequences to make your phone checks less rewarding, which will aid your brain in differentiating between distractions and genuine breaks. For instance, establish a rule that if you check your emails, you must deal with all the emails you encounter.

- Adjust your phone's lock screen during focus times, which will give you an instant reminder each time you tap or lift your phone.

Strategy

The Five-minute Interruption Test

This test helps to (a) build your executive stamina and train your brain not to act on impulses immediately, and (b) determine whether the urge to self-interrupt is a habit or a need for a brief disconnection.

How it works:

- When you feel the need to self-interrupt, dedicate those *five* minutes to persisting with your current task or to practising the Surf the Urge technique (see page 52), maintaining an acute awareness of your internal emotional shifts.

The test:

- If the urge to self-interrupt fades during the five-minute period and you are able to refocus, this was likely an automatic response.
- If five minutes have elapsed and you still want to self-interrupt, take a strategic and intentional break where you do something restorative.

If you wanted to self-interrupt with:

- Small admin tasks - consider inserting them into your scheduled admin block, or noting them down manually to complete in a batch.
- Something that you find rewarding - note it down as something to look forward to (Regulate Rewards - Building Block 7) in your next break.

REDUCE EXTERNAL INTERRUPTIONS

- Communicate to your co-workers specific times that you will do complex work and wish not to be disturbed. This is because constant external interruptions can be fatiguing and increase self-interruptions. Setting a regular pattern for this will be the start of building good habits.

- Block those times out in your work calendar and set automatic email replies if needed.

BATCH CHECKING

- Consider checking emails and notifications in batches as multiple small interruptions have the potential to be more disruptive than a longer, designated period allocated to using your phone.

- One study has suggested that three batches per day provides an optimal balance: it minimizes constant interruptions, ensures access to vital information, and doesn't trigger excessive anxiety linked to fear of missing out.[75] However, remember to tailor the number of batches to suit your specific situation.

- Turn off notifications where possible, mute instant messaging and do not keep your email client running in the background.

Maximize Your Breaks

	WHAT	WHY
Before	Make them intentional	This means that there is no guilt associated with them.
	Don't wait until you are completely exhausted	Once you are in low power mode, it will take even longer to recover.
	Time them right to produce artificial deadlines	Align your autopilot with your executive: use this synergy to complete a burst of work prior to a break.
	Look forward to them	Anticipation increases dopamine. This trains your brain to look forwards to rewards rather than wanting to obtain them immediately.
During	Choose your activity carefully	Opt for activities that energize rather than deplete you. To do that you will need to pay close attention to how you feel afterwards, as the effects will be personal to you.
	Head outdoors	Natural light can have antidepressant properties by increasing dopamine release, which can make your breaks more rewarding.

10

Sleep

It is not uncommon to think of being asleep as the opposite of being awake, with the brain having a binary function mimicking that of a light switch. But sleep is far from a simple 'off' switch. Our brain during sleep can be more dynamic than when we're fully awake, humming with activity across certain regions. While we may recognize how a solid night's sleep – or the lack thereof – affects us, we often undervalue the remarkable behind-the-scenes restoration process that sleep provides to keep us functioning at our peak.

Sleep deprivation, even in the short term, can impact our cognitive abilities, undermining our focus, slowing reaction speed, and hampering our ability to think clearly.[76] Sleep is also crucial for long-term memory and learning; it's during sleep that memories move from our hippocampus, our brain's memory inbox, to various other regions for permanent storage. This process, known as memory consolidation, functions optimally during sleep, when we're not constantly being flooded with new information. Thus, relentless daytime work coupled with inadequate sleep at night, far from promoting productivity, actually works to our disadvantage.

Rather than thinking about sleep as a luxury or an indulgence, it is best to think of it as a time in which the brain performs essential maintenance. Think of sleep like a dedicated repair crew that works the night shift. While the rest of the world is quiet, your brain is bustling, tidying up the mess from the day and getting everything ready for a fresh start in the morning. Sleep gives the brain an opportunity to clear waste products,[77] some of which can be toxic when they accumulate, such as beta-amyloid which contributes to Alzheimer's disease. Our brain is the control centre of our body and not performing this essential sleep maintenance can affect our physical health. Sustained lack of sleep predisposes us to a broad range of health issues such as increased heart disease, lowered immunity, and a higher risk of cancer among others.

Sleep plays a crucial role in our ability to sustain focus and exert willpower. Its restorative functions rejuvenate our executive brain, effectively recharging our 'executive battery'. Consequently, we awaken each day with a fresh supply of executive power after a good night's sleep. Since the executive brain is also essential for emotion regulation, quality sleep contributes to mental toughness[24] – a term scientists use to describe our ability to positively face challenges, deal with controllable factors confidently, remain resilient, and view obstacles as opportunities for personal growth. Conversely, poor sleep quality can lead to a higher propensity for negative thinking[78] which has the potential to affect our mental health. There are numerous studies that show that improving sleep has mental health benefits.[79] One of the main motivations for writing this book was my realization that, aside from sleep, phones represent our most frequent time-consuming activity, and often, the use of our devices encroaches on our sleep.

Our phones often shoulder the blame when it comes to sleep issues and we've learnt that if we don't sufficiently replenish our executive through adequate rest, we are in danger of building and relying on negative digital habits. Yet the interaction between sleep and our phones isn't as simple as we are often led to believe. So, this begs the question: How are sleep and our phone habits intertwined, and is the whole picture truly as bleak as it's painted? Let's dive deep into the fascinating world of sleep and see exactly what part our digital companions play in our nightly rest.

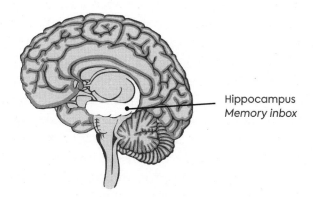

Hippocampus
Memory inbox

The Memory Inbox: Sleep is crucial for memory consolidation, the process by which memories are transferred from the hippocampus, the memory inbox, to be distributed throughout the brain for permanent storage.

The master clock

Inside all of our brains, there is a 'master clock'. Despite measuring only a few millimetres in size, this clock – the suprachiasmatic nucleus – wields a significant impact on our body's daily rhythm. It contains genes, aptly named the 'clock genes', that provide the instructions needed to manufacture hormones which determine how alert or sleepy we feel. This produces a twenty-four-hour pattern of alertness and sleepiness that is referred to as our circadian rhythm. The two key hormones that determine our circadian rhythm are cortisol and melatonin. Cortisol, released by our adrenal glands, peaks in the morning to make us feel alert, and melatonin, produced by the pineal gland in the brain, is released at night to make us feel sleepy.

Much like our ability to focus, genetics play a big part in how this clock functions. All our biological clocks have a natural predisposition to certain waking and sleeping times which in turn affects the timing of hormone release. Our innate predisposition is referred to as our chronotype. Family and twin studies, as well as large-scale genetic analyses, have shown us that a large portion of our chronotype is genetically determined.[80] (It is polygenic, meaning that rather than a single gene, multiple genes have an influence on our master clock.) This means that we all have a natural tendency of when, if left to our own devices, we would prefer to go to sleep and to wake up. Early chronotypes or 'morning larks' awaken early and feel most alert in the morning whereas late chronotypes or 'night owls' awaken later and feel more alert in the late evening or at night. Some of you, while reading this, may immediately strongly identify with an early or a late chronotype but not all of you will. Chronotype is a continuum and the binary division into morning larks and night owls is akin to labelling everyone's height as either tall or short, with all the confusion and difficulty that would ensue. As with height, epidemiological studies show that the chronotype spectrum follows a bell-shaped curve with most peopled clustered around the average and smaller numbers at the extreme ends. So, if you do not strongly identify with being an early or a late bird, you are most likely an intermediate bird, who may have mild early or late tendencies depending on which side of the bell you are closest to.

Our chronotype also depends on our age. On average, children are predisposed towards having early chronotypes. During the teenage years, their chronotype moves later, reaching a maximum lateness around age twenty. This going to bed late and getting up late is not laziness but is, in fact, down to hormone secretion in the brain: in a teenager, the hormones required to go to sleep are released much later than someone who is older. After the teenage years, our chronotype moves earlier, something that is probably accelerated in those of us who have to care for young children who typically get up early. Over the age of sixty, on average, we will become even earlier chronotypes than we were as children.[81]

The power of light

Our genetics and age have a big impact on our master clock but there is nothing that we can do to alter them. They are what is referred to as non-modifiable factors. But there is another factor that has a huge sway on our master clock and which we have more control over: light. When light enters our eyes, it is detected by specialized photoreceptors located in the retina. These photoreceptors generate signals that travel along each of the optic nerves at the back of our eyes, to the junction where they connect and cross over – this is called the optic chiasm. Just above this point lies the suprachiasmatic nucleus, therefore allowing our master clock to receive critical information about the amount of light in our surroundings. It then uses this to regulate our sleep-wake cycle by timing the release of cortisol and melatonin.

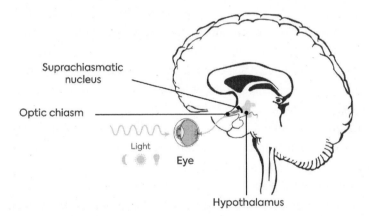

Light Pathway to Eye/Brain: Light hits the back of our eyes and is transmitted via the optic nerves to the suprachiasmatic nucleus, our master clock. This master clock gives instructions to produce the alertness hormone cortisol or the sleepiness hormone melatonin.

When it comes to the impact of light on our master clock, there are two elements to consider: dose and timing. Dose refers to the amount of light our brain receives and is measured in lux. Outdoor light is the most powerful, with exposure to direct sunlight providing a whopping 100,000 lux. Even being in the shade on a sunny day provides 10,000–20,000 lux. On a cloudy day, outdoor light may drop to around 5,000 lux, but this is still significantly more than the 200–400 lux typically found indoors, which dramatically reduces the dose of light our brain receives. The brightness of our indoor environment depends on several factors, such as the number and proximity of windows and the strength of our artificial lighting but, as a general rule, the dose of light that we get inside is a fraction of what we can get outdoors. In fact, even on the dreariest of days, being outside will provide us with at least ten times the dose of daylight compared to being inside.

When we expose ourselves to light, it helps synchronize our master clock with our surroundings, and the timing of light exposure determines the nature of this effect. Exposure to light early in the day will shift our sleep-wake cycle forward, meaning that we wake up and go to sleep earlier. Conversely, exposure to light later in the day will shift our sleep-wake cycle back so we fall asleep later and, even if we set an alarm clock to wake early, we will not feel alert till later on. It's important to note that light exposure at the extremes of the day has a more powerful effect on the master clock than light exposure in the middle of the day which has little to no effect.

Shifting your master clock by a few hours is possible with strategic light exposure, but your chronotype and age will nevertheless determine the starting point. That being said, we should move away from the idea that a particular chronotype is preferable or that being a late riser is some sort of personal flaw. After all, it's largely down to genetics and age. In general, modern society vastly favours morning chronotypes when it comes to school and work schedules, or concepts of when it is best to exercise, and it is too easy to forget that having a range of chronotypes has been a historical advantage when it comes to our safety. Evening chronotypes were able to remain vigilant at times when morning chronotypes would be asleep, and there are still great benefits to having different chronotypes in certain lines of work or when looking after babies.

Visible light comprises different wavelengths which produce the various colours that we see. Long wavelengths produce red, orange and yellow while short wavelengths produce blue, indigo and violet. The specialized photoreceptors in our eyes that detect light contain a chemical called melanopsin (not to be confused with melatonin) which is more sensitive to shorter wavelengths, such as blue light, compared to the longer wavelengths of orange/ yellow light, and will transmit that information to our master clock accordingly. Our devices emit a high proportion of blue light which has received a lot of media attention and blame for the effect that technology has on our sleep. One of the main worries is that blue light can suppress the hormone melatonin which is needed to kickstart the sleep process. Technically, one could conclude that the blue light emitted from phones, tablets and computers is responsible for delaying our sleep and confusing our master clock, and many people believe this. However, this story is incomplete and much more complex.

A lot of the experiments around blue light and sleep were done on mice and there is one key difference between mice and humans which we must not overlook: mice are nocturnal animals whereas humans are not. As mice are mainly active at night, the photoreceptors in their eyes are very sensitive to even the smallest amount of light, as low as 0.1 lux, an intensity that is completely ineffective on humans. This is because we have evolved to be active during the day and spend long periods of time outside, typically exposed to light intensity in the region of thousands of lux. Screens will only emit light in the region of 50 lux and our photoreceptors are relatively insensitive to this low level of light emission, even if a high proportion of that is blue light.[82]

The best-known study that is cited as evidence that light from our devices affects our sleep compared two different modes of reading before bedtime. Twelve people were asked to read using either an eBook, which had a blue light-emitting screen, or a physical book.[83] They were randomly assigned to one method of reading for five consecutive days before switching to the alternative method for another five consecutive days. The researchers closely monitored their sleep pattern and found that, when using the eBook, participants fell asleep later than when they read a physical book. As a result, this study has been widely publicized as evidence that technology impacts our sleep negatively and has been used to vilify using our devices at bedtime. However, the findings are somewhat taken out of context. What is less commonly known is how small the difference between the use of the eBook and physical book readers was. Participants using the 31 lux e-Reader fell asleep only ten minutes later than participants reading the 0.1 lux physical book. Total sleep time and sleep efficiency was unchanged. While this effect may be significant in terms of statistics, something scientists get very excited about, it means very little in real-life terms as the difference is too small to notice. Essentially, what this means is that, while technology emits a high proportion of blue light, the intensity that is transmitted is not enough to dramatically shift our master clock.

Therefore, the popular belief that blue light emitted from electronic devices strongly affects our sleep quality is not as robust as is often portrayed. In yet another study, when researchers asked participants to look at a tablet at full brightness for an entire hour and measured melatonin levels, they found no significant suppression.[84] To get a significant melatonin reduction, participants had to wear glasses mounted with blue-light emitting LEDs to deliver extra blue light. You probably need about two hours of exposure to a phone or tablet to get any significant melatonin suppression (again with a screen at full brightness)[85] but, even when that happens, the effect is relatively small and short-lived, with melatonin levels recovering within fifteen minutes of cessation of light exposure.[85]

Concerns over the impact of blue light have led to several technological solutions that seek to reduce the amount emitted from our screens. Smartphone designers have created night-time modes to reduce brightness and blue light, which can be automatically set for certain times. In reality, it is likely minimizing what is already a small effect. For example, one study found that there were no clinically significant sleep differences, including to sleep quality, when participants used a bright tablet screen compared to one that was dimmed or had a blue light filtering program applied.[86] Thus, if you regularly use your phone before bed but have no trouble falling asleep, there is no need to panic that blue light is inadvertently harming you without your knowledge. This is further emphasized by randomized control trials which show that blue-light-blocking glasses have no effect on the sleep of healthy volunteers.[87]*

Rather than hastily attributing sleep-wake cycle disruptions to our smartphones, we might be overlooking a more significant factor: the influence of artificial light within our homes. This was illustrated in a study involving two hunter-gatherer communities in the Argentinean Chaco. Despite their proximity – only 50 km apart – one community had twenty-four-hour access to electricity, while the other community was dependent solely on natural light. Data collected from wrist activity monitors and daily bedtime diaries revealed a significant difference in the sleep patterns of the two communities. The group with access to electricity went to bed significantly later and therefore slept for less time – forty-three minutes less in the summer and fifty-six minutes less in the winter.[90] This is a clinically significant finding and much more notable than the minor ten-minute delay in sleep onset observed in the eBook study. This minimal delay sleep is likely due to the eBook study's unique conditions – participants had to maintain a strict 10 pm–6 am sleep/wake schedule and were strictly confined to a dimly lit room for four hours before bedtime, a scenario unlikely to occur in the typical home with electricity. A different study found no such delay in falling asleep regardless of whether participants read a book or used an iPad emitting a modest

---

* While the impact of blue light is likely insignificant for healthy individuals, it is possible that this effect is increased in those suffering from insomnia, mood disturbance or other neurological conditions, as randomized controlled trials do show some benefits to wearing blue-light-blocking glasses.[88,89]

brightness of 58 lux. Rather than a dimly lit room in a lab, this was done in the participants' own homes, and not only was the minor delay in sleep onset not observed, but also the duration of sleep and the time spent in various sleep stages – measured by placing electrodes on the participants' heads – were unchanged.[91] This implies that our routine exposure to more intense artificial lights in the home is a more significant disruptor of our sleep patterns than our pre-bedtime use of an iPad.

So, when it comes to sleep, rather than fixating on our phones, we need to look at the bigger picture. Artificial light is able to delay our sleep, but our master clock also relies on the levels provided by outdoor daylight to synchronize our circadian rhythm. Being in natural daylight means that we would typically be exposed to light intensity in the order of thousands of lux, whereas today we now spend the majority of the day inside and exposed to just a few hundred. Even on a cloudy day, natural daylight has a powerful effect on our brain, stimulating the production of the alertness hormone cortisol, in a way that our dim brick caves do not. Our master clock which is used to having a clear distinction between day and night (from thousands of lux in the day down to less than 10 lux at night) now has to deal with a lifestyle where the amount of light we are exposed to is practically equivalent. In winter, when the days are shorter, this is even more acute. People may get up early, commute in the dark and then spend their entire day in a building that provides a fraction of the light intensity compared to outside. At the end of the working day, exposure to outdoor light at dusk, which is in the region of 3,000 lux, confuses our master clock even further – especially if we've had little light exposure during the day. As a result, our circadian rhythm becomes even more disrupted, and we feel sleepy during the day and paradoxically alert at night, hence going to sleep later.

In the evenings, artificial lighting in our homes will pose more of a problem than our screens. The tablets and e-Readers tested in the above studies provide a light intensity of 80 lux at full brightness, which falls to under 50 lux when dimmed. Depending on how bright the lights in your home are, they probably provide around 200–400 lux.* Therefore the effects of modern living and working, coupled with artificial lighting at home, have a more powerful effect on our master clock than the blue light emitted by technology. To put this in context, the bright lights in our bathrooms when we get ready for bed will have a bigger impact in delaying sleep than a dimly lit phone screen. It is no wonder that our circadian rhythm becomes dysfunctional and we have problems with our sleep.

Blue light from our screens is not the insurmountable problem it has been made out to be and the commonly suggested advice to dim your phone screen or wear blue-light-blocking glasses is not a panacea for poor sleep. Getting a good night's sleep in fact begins in the morning by using natural light to inform our master clock that the day has begun. To address sleep-wake cycle dysfunction, we need to start by building habits of strategic daylight exposure. The strong signal provided by outdoor light helps synchronize our master clock with the beginning of the day, allowing it to discern the difference between day and night once again. Setting your screen to dim automatically at a certain time is easy to do and might make your phone less rewarding. It might also provide a psychological reminder to the autopilot brain to start winding down. But, ultimately, its effect is limited if you are still exposing yourself to bright artificial lights.

Bedtime procrastination

While the effect of blue light from our screens has been vastly overstated, this does not mean that the stimulating nature of technology never disturbs our sleep. Several studies show an association between using smartphones and going to bed later.[92] However, this outcome is not inevitable; it largely depends on our

---

* All the light intensity values provided here are just a rough estimate but if you are interested in the light intensity in your own environment and technology you can either purchase a low-cost lux meter or even download an app on your phone.

habits and, as usual, we need to delve deeper. First, I want you to consider the following question: have you ever stayed up late despite being tired and knowing that a lack of sleep will have a negative impact on your tomorrow? Bedtime procrastination is failing to go to bed at our intended time despite having no external circumstances that prevent us from doing so.[93] We stay up doing things that are not necessary, delaying our sleep in favour of more enticing activities which can be non-technological but, in many cases, can involve scrolling and binge watching. Bedtime procrastination has three key features: 1) a delay, 2) the absence of a valid reason and 3) the knowledge that our actions will have negative consequences, i.e. we will feel more tired the next day.

So, if we know we will be worse off the next day, why do we do it? Just like other types of procrastination discussed in Chapter 9, sleep procrastination arises due to a conflict between the autopilot and the executive. A number of studies have linked bedtime procrastination to low levels of self-control and reduced willpower.[94] The concept of low power mode, which we previously unpacked in this book, resurfaces as a key player when we delve into our sleep patterns. You see, heading to bed is an inevitable end-of-day task, something we have to do at a time when our executive brain is typically exhausted from the day's events. We've spent the day using our executive function to accomplish challenging tasks, and as a result, we enter this state of mental fatigue. The executive brain running on empty means that the autopilot brain steps in and, acting in low power mode, means that we make choices based on instant gratification, our executive brain too exhausted to consider the delayed negative consequences that we will face the next day.

Sometimes bedtime procrastination is an active choice, a response to a difficult day. If you've not had a chance to take a break, it may have become a way to reclaim 'me-time'. Other times, bedtime procrastination is a passive choice where we get absorbed in an activity and don't realize how much time has passed. We may procrastinate prior to going to bed or while being physically in bed. If you can relate, you will instinctively know which of these scenarios applies to you – it is not uncommon for there to be a combination of both. A lot of bedtime procrastination is down to habit[95] because, when this behaviour is repeated over time, bedtime procrastination scripts are coded into our autopilot which make it progressively easier to engage in this behaviour.

When we delay going to bed because we are on our phones, it is not the phone itself or the type of light emitted from the screen, but the content we access that has a greater influence on falling asleep. Not all things that we do on our phone are equal and certain types of content have a greater detrimental impact on our sleep than others. For example, listening to a calming story is not the same as having an argument with a stranger on social media. For most of us, being tucked in bed in the confines of our own homes signifies safety. Our cortisol levels should be waning, and we should be feeling sleepy. Cortisol, the hormone which is usually released in the morning to make us feel alert, is also released during times of stress. This is a strong evolutionary safety mechanism to ensure that, instead of feeling sleepy, we become attentive and vigilant at times of potential danger. The benefit of this is something I've personally experienced – for instance, if I'm awakened in the middle of the night for an emergency at the hospital, I find myself gaining focus faster than if I were woken up for any other reason. This alerting effect can be so powerful that it persists for several hours afterwards, making me unable to return to sleep.

Viewing anxiety-inducing content prior to our bedtime will signify danger to our brain. Browsing the news means we will be exposed to threatening global events or cases of individual harm. The more shocking an event, the more likely it is to be written about. We may be privy to an argument on social media. Within our inbox, we may find an email that causes us to worry. Our brain, unable to discern the difference between our immediate physical environment and the online world, releases cortisol as a result of being exposed to potentially upsetting content. It decides to remain vigilant. This feeling of alertness means that we are even less likely to want to go to sleep and our sleep-wake cycle is significantly delayed as a result.

Stressful situations are not the only triggers of our alertness. Social interactions similarly engage our attention and interestingly, they're factored into our biological rhythms. With the bulk of social interactions happening during the day and easing off at night, our brain uses this pause to instigate our sleep. There is a huge variation in how we use our phones before bed, but it is thought that socially interacting with others (e.g. texting friends) has a greater psychologically stimulating effect and delays our sleep more compared to doing something passive such as watching a video.[92] Notably, these effects are not restricted to our phones and having an in-person argument in the evening, something which is both stressful and a social interaction, will also impact our sleep negatively.

However, while stressful content and social interactions have specifically alerting effects, any content that we become engrossed in can increase our alertness at a time when we should be powering down. Again, this is not unique to our phones and can happen with other non-technological activities. I've personally stayed up till the early hours to finish reading a book – a great example that bedtime procrastination can affect everyone, including neuroscientists! That being said, the breadth of the content that we can access through our phones as well as the lack of stopping reminders in apps makes it more likely to get absorbed in, and procrastinate with, our devices. In the three hours prior to bed, people who sleep procrastinate spend four times as long on their smartphones compared to those who do not sleep procrastinate, a difference which amounts to sixty-one minutes.[96] In addition to spending more time on their phone in the evening, people who are prone to bedtime procrastination spend more time on their smartphone during the day in general, indicating that they are more likely to have multiple ingrained digital habits, bedtime procrastination being just one aspect of them.

Although sleep procrastination can affect everybody, if you are a natural evening chronotype you will be more susceptible to it.[94] Night owls tend to be more alert in the evenings and may even experience a burst of energy around bedtime. It is likely they are also more sensitive to evening light exposure as well as psychologically stimulating content. Parents are often frustrated when their teenage children stay up till the early hours of the morning, making them tired the next day, but this is not a new phenomenon created by the digital era. As discussed, our chronotype moves later during adolescence, meaning teenagers find it difficult to go to bed as early as adults. A subset of teenagers are more severely affected by this, potentially suffering from sleep delayed phase disorder, characterized by an inability to fall asleep until between 1 and 6 am. Under these circumstances, spending time on smartphones often becomes the go-to activity for teenagers. This fills the unsocial hours while they wait for their brain to release melatonin and induce sleepiness. Unfortunately, the stimulating content on their phones, coupled with losing track of time, exacerbates the problem, leading them to scroll for longer than they initially intended. As a result, they will stay in bed later and this pattern, coupled with reduced morning daylight exposure and increased evening artificial light exposure, results in them feeling even less sleepy the following evening. This could potentially kick-start a vicious cycle, where bedtimes are progressively delayed. In this cycle, phone habits emerge as one of the many contributing factors to sleep disturbances – a factor which, with the help of the techniques discussed in this book, can be effectively altered.

Digital sleep aids

A commonly touted piece of advice is not to check your phone for a set number of hours before bed which, unsurprisingly, some find hard to do. Why is this so difficult? Executing an all-or-nothing rule is particularly difficult if you are exhausted at the end of the day and running in low power mode. Going to sleep is not like turning off a light switch – not just because the brain remains active but also because going to sleep is not as simple as pressing a button. Sleep depends on several subconscious areas of the brain, and is an involuntary process, something that people who have difficulty sleeping, no matter how hard they try, will attest to. One of the biggest mistakes I ever made as a medical student was that I would study all day and then go to bed the minute I'd finished. I thought that winding down was a waste of time. But I soon learned my lesson as I lay in bed for hours despite being exhausted. This is because going to sleep is like trying to stop a moving car. The faster it's going, the longer it will take to stop. So, if you've had a particularly hard and emotionally charged day, your brain will be racing and you will therefore need to spend some time processing events and changing gears to enter a mode where you can fall asleep.

Our brain is a powerful association machine and uses the information provided in our environment to get into certain states. I've previously discussed athletes' individual warm-up routines which prepare them physically and mentally before a match, and this is something I recommended that you implement prior to a focused work session. Sleep is no different and we have all developed our own personal sleep routines that prepare our body and mind for relaxation. Many people reach for devices in the evening as an easy way of winding down and, over time, our brain will identify the use of digital devices as a sign to begin the relaxation process. They become a digital sleep aid of sorts. Yet this is not necessarily a bad thing. Simple tasks such as playing games or passively watching content are activities that can be done on autopilot. They require little effort from the executive hence giving that part of the brain, exhausted from the day's events, a chance to relax and recover. It is no different to the escapism that reading a book provides at bedtime. Ruminating about work or other events that happened during the day can be a source of stress which impacts sleep quality. This increases if you find yourself in low power mode and therefore have less internal resources to process your emotions. Engaging in low-effort activities on smartphones therefore provides a means to counter this and disconnect from work-related activities and thoughts.

Whereas using our devices for sleep procrastination has a negative effect on both sleep duration and quality, using our devices as a form of psychological detachment has been shown to have a positive impact.[97] It is important to note that these two factors, bedtime procrastination and psychological detachment, are not mutually exclusive. Some people may engage in one of these behaviours, but most will find that they do a combination of both. Sometimes, people may rely on their devices to wind down but concurrently sleep procrastinate to an extent that, rather than actively going to sleep, they will wait until the degree of exhaustion is so great that they fall asleep with a video on their phone still running.

Instead of saying 'absolutely never use your phone' – an impractical and undesirable notion for many – I propose a different approach: if you want to use technology close to bedtime, do so in a more mindful way. Given how common it is to use smartphones in the evenings, many people will be relieved to hear this. Rather than a total ban, we can manage our digital interactions to limit adverse effects on sleep. The key lies in replacing harmful digital habits with beneficial ones (as discussed in Building Block 9). It's all about choosing your content wisely. For instance, if reading Twitter debates makes your blood boil, consider switching that habit with reading something less emotionally provocative. Similarly, if an enthralling TV series keeps you up, try replacing it with calming sleep-inducing stories. Even when it comes to gaming, opt for games with less novelty and built-in stopping reminders, which can help you wind down without overstimulating your mind. Remember, we all have different ways to unwind; what one finds relaxing may be stimulating for another. Therefore, through some trial and error, find what works best for you. In addition, Leveraging Location (Building Block 5) can prove to be instrumental. Try setting boundaries for the type of content you allow in your relaxation and sleep zones. We do not want to associate our bed with anything stressful or stimulating. With this knowledge in mind, it is up to you to develop your own guidelines, remembering that these must be realistic enough to achieve the necessary repetition to establish new habits.

Understanding the potential pitfalls of bedtime procrastination empowers us to make informed decisions about our device usage. Our aim is not to vilify devices but to use them in a way that serves us best. Doing so in a mindful way appears to be the key to maximizing the positive benefits of psychological detachment and protecting against negative effects, ultimately enabling technology to work for us, not against our sleep.[97] We'll delve further into the role of mindfulness in the next chapter, focusing on mental health.

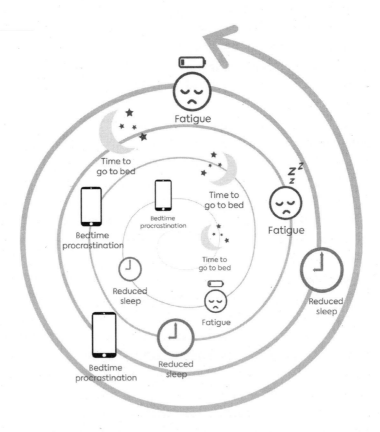

The Sleep Procrastination Cycle: We often put off sleep because we're in lower mode, marked by diminished self-control. But, here's the catch: sleep is essential for replenishing that self-control. Delaying sleep only leads us to feel more depleted the next night, perpetuating a vicious cycle of further sleep procrastination.

The sleep stopwatch

It is not an unusual scenario to spend a considerable amount of time in bed scrolling before we get up in the morning. This may have started off as a form of digital procrastination, a sign that our exhausted brain is putting off starting the day by delaying getting out of bed. Alternatively, browsing social media, the news, or our email is a solution of sorts, as we use the cortisol-stimulating nature of this content in a desperate effort to feel alert following insufficient sleep. Over time, the action of checking our phone straight after we wake up has become stored into the autopilot brain as a habit, an integral part of our routine and something we do without thinking.

Waking up feeling fatigued means that we are more likely to engage in the dreaded early-morning scroll and also less likely to have the capacity to change our behaviour. But the underlying reasons why we feel so depleted have actually started prior to that morning. Our brain needs a consistent schedule, something that dawn and dusk provide, to ensure the master clock remains in sync with the solar day. This is because the process of signalling from the master clock to the adrenal gland to make cortisol and make us feel alert is not immediate. It takes time to get the messages across to produce these hormones. This means that, rather than responding to our needs, this system anticipates them. For maximum efficiency, our brain starts to signal for the production of cortisol several hours before we wake up, already expecting our need to feel alert. Having a regular sleep-wake cycle is therefore crucial for the brain to predict when to produce the required hormones to make you feel sleepy or awake.

Our master clock can slowly adapt to seasonal variation – our sleep also needs to change, meaning that we sleep more in winter and less in the summer – but it cannot adapt to large abrupt changes in schedule. Therefore, getting up at different times each day confuses this system and causes our hormonal production to be out of sync. Suddenly waking up several hours earlier means that there will be a lag until cortisol is produced and we feel alert. This lag is similar to jet lag which occurs during travel across time zones and means that our own master clock becomes out of sync with our environment until we adjust to our new time zone. Except, rather than the occasional flight, erratic sleep-wake cycles can occur on a regular, almost weekly basis. This is known as 'social jet lag', a phenomenon that typically results from a sharp shift in our sleep pattern from the weekend, where the typical person gets up later, to the weekday, when we generally have to get up early.

A rapid adjustment to an earlier waking time at the beginning of the working week impacts many people, with evening chronotypes most greatly affected. Having to get up early then means that the duration of our sleep is shortened. On days off, we stay in bed late to catch up on sleep, meaning that our waking and sleep times shift progressively later. This effect is exacerbated further as staying in bed later means we get less exposure to early morning natural light.[98] This causes our sleep pattern to become erratic and has a knock-on effect on the brain's master clock. Shift workers form a group that is commonly affected by this due to their frequently changing shift patterns, and studies show that their cortisol production is disrupted.[99]

Our brain possesses an internal stopwatch that measures how long we have been awake, using a small chemical molecule called adenosine. Adenosine builds up in the brain during the day, measuring how long we have been awake, and is then cleared during sleep. This build-up of adenosine also increases tiredness levels and puts pressure on the brain to sleep, something referred to by neuroscientists as sleep pressure.

Adenosine requires a good night's sleep to be fully cleared – this essentially resets the sleep stopwatch. However, if you do not get sufficient sleep, you will start the morning with a higher adenosine level meaning you will wake up feeling tired. Adenosine then continues to build up during the day, so if you've had a single bad night's sleep, the extra adenosine will put increased sleep pressure on your brain and you'll feel more tired and sleepy earlier, to try and make up for your lost night's sleep. Being continually under-slept leads to an accumulation of adenosine and, hence, persistent levels of fatigue.

Feeling too fatigued to start the day means you may procrastinate getting out of bed and reach for your phone, handily located on your bedside table. An early-morning scroll is not necessarily problematic in itself, but it stops us doing the very thing that would help our master clock: getting exposure to natural daylight. Shortly after dawn, outdoor light is at least thousands of times brighter than lying in a dark bedroom and staring at a dimly lit screen. Natural daylight doesn't just fine-tune our master clock; it provides a host of other brain benefits. For instance, a large-scale UK Biobank study involving half a million people discovered that for each hour spent outdoors, the frequency of tiredness and symptoms of insomnia was noticeably reduced.[100] Additionally, it reported a positive impact on mood, with participants demonstrating lower likelihood of experiencing a major depressive disorder. It is believed that these antidepressant properties of light occur due to a decrease in activity in a part of the brain known as the lateral habenula.[101] This region typically suppresses the ventral tegmental area, a major dopamine producer that we discussed in Chapter 7. Outdoor light therefore releases the lateral habenula 'brake' on this region leading to increased dopamine production. As we have learned, the two key functions of dopamine are to provide a strong learning signal, and subsequent motivation. Prioritizing some light exposure before your early morning scroll is a good method to reprogram a problematic early-morning scroll habit. To put this in action, set implementation intentions (Building Block 4) to apply the Five-minute Rule (Building Block 1) and use that time to expose yourself to natural light or, at least, open your curtains as a Plan B (Building Block 2) even if you get back into bed to check your phone. To hold yourself even more accountable, move your phone from your bedside table next to a window as a reminder and another form of Pre-commitment (Building Block 4).

The most effective way to reverse adenosine build-up is through adequate, quality sleep which is aligned to our natural circadian rhythms. But, unable to do this, in practice we commonly reach for the world's most popular psychoactive drug instead. Caffeine works by blocking the adenosine receptors in the brain. The internal stopwatch is still running, adenosine will continue to build up, but the caffeine consumed is stopping your brain from seeing it. When the caffeine is eventually metabolized by the liver, it creates an energy slump, the effects of which may be felt acutely, but it may also have longer-lasting effects that are not immediately obvious.

A comprehensive meta-analysis conducted in 2023 pooled data from numerous studies and found that caffeine consumption led a decrease in sleep duration by 45.3 minutes.[102] Even if falling asleep after consuming caffeine isn't a challenge for you, the efficiency of your sleep could be compromised as the study also found that caffeine intake led to significantly more light sleep and reduced deep sleep – the phase when the body usually clears out adenosine – corresponding to a 7 per cent reduction in sleep efficiency. In essence, we may find ourselves in a situation where we try to counteract rising adenosine levels using caffeine but, in doing so, we are not just masking the problem but also inadvertently making it worse. We further reduce our clearance of adenosine by both decreasing sleep time and reducing efficiency, and then find ourselves in a perpetuating cycle of battling raised adenosine levels and trying to self-medicate via escalating doses of caffeine.

This is not to say you must never use caffeine but, consistent with the key theme of this book, be well-informed about its effects on your brain. Much like our smartphones, caffeine isn't inherently bad – it's how we use it that counts. Specifically, the timing and dose make a huge difference. On average, your liver takes about three to five hours to break down half of the caffeine you consume – that's what scientists call the 'half-life'.* The study

---

* Scientists often use the term 'half-life' when referring to the metabolism of substances because it provides a more reliable measurement. When the concentration of a substance dwindles in the system, the rate of breakdown by enzymes also decreases. This is analogous to a checkout clerk who works at an efficient pace when there's a queue of customers but slows down when there are fewer shoppers, interspersed with pauses when no one is waiting.

found that for caffeine not to adversely affect sleep, and to allow the caffeine to be metabolized, there should be a cut-off of 8.8 hours prior to sleep for a cup of coffee (which contains around 107 mg of caffeine) and 13.2 hours for pre-workout drinks, which have higher caffeine content (217 mg). So, if you're aiming for a 10 pm bedtime, this translates to stopping drinking coffee at around 1:12 pm. Tea, with its lower caffeine content (47 mg), didn't significantly impact sleep, so you could, theoretically, enjoy a cup closer to bedtime.

Bear in mind that these guidelines are averages, and much like the one-size-fits-all concept doesn't apply to a pair of jeans, it doesn't work for caffeine metabolism either. We're all different, and there's a substantial genetic variation that influences how our liver enzymes process caffeine. Some of us metabolize caffeine quickly, hence are less sensitive to its effects, while others process it more slowly and are more sensitive. That's why the range of caffeine metabolism is probably broader, more like two to eight hours. So, if you find yourself constantly tired and your sleep isn't as refreshing as it should be, swapping your caffeinated beverages with decaffeinated ones (which contain a mere 2 mg of caffeine) could be a straightforward solution to experiment with. As with any change, observe how it affects you, adjust, to find a strategy that works for you.

So, when we carefully consider how the quality of our sleep relates to our devices, we find that our phones are able to affect our sleep, but we should not blame our sleep disturbance entirely on technology. Rather, our phones represent one among many contributing factors, each of which I've endeavoured to explain in detail throughout this chapter for a balanced perspective. Reduced or poor-quality sleep does not just mean we wake up feeling tired; we also wake up unmotivated and with an undercharged executive battery, putting us in a state where we are more likely to form negative digital habits, some of which in turn have the potential to affect our sleep. The early-morning scroll may reduce exposure to natural daylight and bedtime procrastination may mean we forgo sleep in the evening. The overall result is yet another vicious cycle of ever-reduced sleep which continually exacerbates our negative digital behaviours. Improving sleep is an investment that pays off with increased levels of alertness, focus and willpower and therefore makes tackling any problematic digital habits surrounding our sleep of utmost importance. Not everyone reading this book will want to stop using their digital devices at bedtime, and I think that is OK – part of life is being able to be supported in making informed choices. By understanding how our sleep cycle works and the times when our phone use can help or hinder this – rather that outright fearing its effects – we can strive for a healthier sleep routine.

Sleep Practical

Sleep and Smartphones

- Smartphones and tablets lack the necessary intensity of light to modify our master clock.

- Stressful smartphone content can stimulate cortisol production and have alerting effects.

- Smartphones can provide a means for bedtime procrastination, especially in low power mode.

- Smartphones can offer beneficial psychological detachment when winding down before bed.

- Inadequate sleep can contribute to morning scrolling habits through executive fatigue and increased adenosine levels.

How to optimize your master clock

- Implement regular waking and sleep times where possible. If you struggle with sleep disturbances or have difficulty falling asleep, establishing a consistent wake-up time – known as 'anchoring your morning' – can be a powerful strategy. Unlike falling asleep, which is involuntary, waking up at the same time each day is something that we can control to help regulate our sleep cycle.

- Use exposure to natural light to strategically adjust your sleep cycle – if you want to advance your sleep cycle (i.e. wake up earlier) light exposure should happen in the morning as close to waking up as is reasonably possible. Exposure to natural light in the middle of the day will not alter the master clock but has other benefits such as increased dopamine release.

- If you find you wake up too early and would like to delay your sleep cycle (i.e. go to sleep later and wake up later), invest in blackout curtains, avoid morning light and aim to spend more time outdoors in late afternoon/evening if it's still light. Otherwise, you can use the brightness of your home lighting to keep you alert till later.

- If you do want to sleep in longer at the weekend, try limiting this to one hour (don't worry if this is not possible and you need to catch up as it is better to get more sleep in this situation; however, it is likely to be more important that you focus on improving weekday sleep). Make sure to offset this delayed start through natural daylight exposure as soon as you get up.

Daylight exposure: What you need to know

- Natural daylight is best but, if that is not an option, you can use a light box (most deliver around 10,000 lux and this can be adjusted to lower intensities).

- Outside is always better than being inside. Not being surrounded by brick walls means the intensity of light increases at least tenfold.

- Whether it's a sunny day or cloudy day does not matter. You can always make up for reduced intensity by staying outside for a bit longer.

- Thirty to sixty minutes of daylight is recommended to get maximum benefits but remember the 80/20 Rule which states that you get the greatest benefits going from nothing to something. Even a few minutes will help you build a habit.

- You do not need to be in direct sunlight. You can get the same exposure in the shade as this is about what your eyes can see.

- Wearing sunglasses reduces the effect of daylight exposure - remember we are talking about morning and evening exposure, not peak of the day!

- You can (and should!) wear sunscreen as the clock is synchronized through your eyes, not your skin.

- As a general rule, you should avoid looking at any light that causes pain and discomfort and you should not look directly at the sun.

★ Remember: caffeine has an alerting effect by blocking your brain from seeing its internal stopwatch. Be mindful of your consumption and consume it strategically (i.e. when you need a boost) rather than out of habit. For example, if you have a day off, you may not need to drink caffeine, particularly if you've slept later. Try to switch to decaffeinated drinks in the afternoon - these still contain some caffeine but the amount is too small to have an effect.

Tackling bedtime procrastination

- Make heading to bed fulfilling by pairing it with another rewarding – but also relaxing – activity. For example, choose certain bath or skincare products for bedtime or limit reading a book that you enjoy to the evenings only.

- Make sure to take breaks throughout the day to recharge your executive brain. If the evening is your only chance to have free time you are more likely to procrastinate, particularly if your executive is exhausted.

- Mindfulness at bedtime has been shown to be beneficial and is a skill that can be honed. Allow yourself permission to wind down with your favourite tech activity but remain self-aware of when this slips into bedtime procrastination territory.

- Add some tech-free associations to your wind-down routine as these are more likely to be finite. This is not about removing but adding, so create a large repertoire of things that your brain associates with sleep rather than depending solely on your device for psychological detachment.

Some ideas:

- Hot bath or shower – this has the added benefit that it causes the blood vessels on our skin to dilate (i.e. get bigger) leading to a drop in body temperature which aids sleep.

- Smells – lavender is most often linked to relaxation but the type of scent you use is not important as your brain is able to form sleep associations with any smell when used consistently.

- Journalling – a great way to empty things that your brain is holding onto is to put them onto paper. Use the tips provided in Part II to start a Really Small Habit (Building Block 8).

★ Remember: Sleep procrastination might feel like you are gaining extra hours in your day, but this is akin to a loan - borrowing self-control from tomorrow - and one which your future self will repay with interest the next day.

Better phone use at bedtime

- Dim the brightness of devices and use blue-light filtering tools. Adjust settings so that this is done automatically at bedtime. While this has a minor effect, it is a simple action which can also act as a signal to your autopilot brain that it's time to wind down and makes your phone less rewarding.

- Unpredictable and anxiety-inducing content should be avoided. Engage with something light-hearted instead.

- Passive content is preferable over active content and, in particular, over social interactions.

- Finite activities are better than activities with no stopping points. Examples: playing a game with a natural endpoint, doing a meditation or listening to a sleep story. Create stopping reminders to ensure that you do not get absorbed for too long and suffer reduced sleep duration as a result.

- Notifications that may disturb your sleep should be switched off.

★ **Remember:** Waking up in the middle of the night is a common occurrence for a lot of people. It is not advisable to check your phone at that time, as a portion of your adenosine has been cleared and sleep pressure has reduced. This means that the psychologically alerting effect of looking at something on your phone will disrupt your ability to get back to sleep. If you are tempted to do this, Insert Hurdles (Building Block 3) by turning your phone off or apply Precommitment (Building Block 4) by placing it away from your bed.

Break the early-morning scroll habit

1. DELAY:

- You do not need to have an all-or-nothing approach. Start by delaying your morning scroll using the Five-minute Rule and increase from there.

- For an extra alertness boost, use those five minutes to obtain as much natural light as possible. Open your curtains and, if possible, consider opening your windows or heading outside. Even doing this for a few minutes will be better than looking at a dimly lit phone screen.

2. REDUCE:

- Set yourself a timer of how long you can scroll for and aim to gradually reduce it every few days until you get to a manageable amount. Remember: don't dismiss your timer.

3. SUBSTITUTE:

- Replace your early-morning scroll habit with something else. My favourite is writing three things I am looking forward to in a journal, which harnesses the power of anticipation (Building Block 7) but is also finite. Try exploring books or magazines that feature engaging quotes or affirmations.

Tips:

- Pre-commitment (Building Block 4) helps if your willpower is running low. Rather than charging your phone on your bedside table, move it far enough away that you have to get up to get it. Alternatively, buy an alarm clock and place your phone even further away.

- Habits are reinforced through reward. You can reward yourself with the time saved by having a leisurely breakfast of your liking.

- Keep persevering and don't worry about setbacks. It will take a couple of months for your new morning routine to become automatic.

★ **Remember:** Snoozing is a false economy. The brief timing of it means that your brain does not reach the required sleep states to have a restorative effect. Rather than setting an early alarm to account for snoozing and early morning scrolling, it is more restful to sleep later but get up immediately.

11

Mental Health

The biggest concern regarding smartphones is undoubtedly their impact on our mental health. As we scroll through the news, we are bombarded with headlines alerting us to the dangers of the very device we hold in our hands. And those headlines certainly don't hold back. They range from gentle inquiries – 'Are smartphones contributing to mental health issues?' – to the more unflinchingly direct assertions that 'Phone Addiction is Real – and so are its Mental Health risks'. More often than not, the suggested remedy is as drastic as it is simplistic, and mainly aimed at the next generation: an all-out ban on smartphones.[103-105]

Opinions on the topic are certainly strong and, based on the claims of these sources, you would think that we know the true impact that our devices have on our mental health – and that it must be bad. We've taken for granted that this is true (though it hasn't had much effect on our phone use!). There is a general sense of fear and anxiety about the effects of phone use on our well-being and many of the population's mental health issues are attributed to the increased adoption of smartphones and their apps, social media in particular. But rather than being facts, these are assumptions. Assumptions that can be so ingrained and visceral that we do not think to question them.

The relationship between our phones and our mental and physical well-being is much more complex and nuanced than the straightforward negative reputation assigned to it by newspaper articles and the studies they quote. The first thing to recognize is that the information frequently communicated regarding technology is biased. Part of the purpose of news is to capture our attention and it is much more likely that research studies with negative findings will be deemed newsworthy enough to be reported. Alarmist headlines tend to gain more traction and to be shared more frequently. The research, however, tells another story. Rather than being unambiguously good or bad, studies on phone use and mental health have shown both negative and positive effects – or even no effects at all. Moreover, nearly two decades' worth of research has failed to conclusively show a strong negative effect of phones, or their apps, such as social media, on our well-being.

There are two key reasons for this, which are perfectly logical but get lost in all the noise and hyperbole. The first is that spending time on our phone is not inherently negative. By picking up your phone you are not predestined to harm your mental health because its influence on us depends very much on how it is used. Using your phone to message your friends will not have the same effect as checking your email, reading the news, or looking at social media. But, even within those examples, the effect on your well-being will depend on who you're messaging – are they supportive or unsupportive? – and the type of content you are consuming. Unfortunately, in research, a lot of these activities have been lumped together under the umbrella term of 'screen time', which makes the data so hard to interpret that it ends up being meaningless. More screen time doesn't automatically decrease well-being and there is no sharp cut-off after which your mental health will suffer.

The second key factor depends on the complexity of people. No two human brains are alike. Each of us has our own unique likes, dislikes, strengths and vulnerabilities, some based on our genetics and others influenced by the environment in which we grew up or currently exist in. Studies which show no effect on mental health do not indicate that there was no effect on everybody. Rather, a mix of positive and negative effects ended up cancelling each other out.

A key study that aptly illustrates this monitored 387 young people using different social media platforms over a period of three weeks. The researchers found that 45 per cent experienced no changes to their well-being, 28 per cent experienced negative effects and 26 per cent experienced only positive effects.[106] Lumping everyone in that study together would show either no overall change or a tiny negative effect. The average of a large group says little about each individual participant and, given that you are reading this book as an individual, the following sections will cover what to look out for and the questions to ask yourself in order to determine whether your phone is having a positive or a negative impact on your mental health. This is not – and can never be – an exhaustive guide to mental health in the online world, nor does it offer a quick-fix solution to any issues you might be experiencing. However, I hope it will give you some tools to think about your mental health and digital behaviours in a critical way, and to use your own intuition about your relationship with your devices, outside of the narrative that is portrayed in the media.

Before I continue, it is important to say that if you feel that you are suffering from a mental health illness you should seek personalized advice from a doctor. In particular, watching self-harm content or being a victim of cyberbullying have been shown to be especially harmful.[107-109] These have a very real and significant toll on your mental health so, if that applies to you or someone you know, it is important to reach out for help.

The emotional brain

To better understand mental health, it is important to know about how our brain processes emotions. When a potentially worrying situation arises, two marble-sized areas near the centre of our brain, one on either side of the midline, are activated. They are called the amygdala, their name arising from the Greek word for almonds due to their appearance. Activation of the amygdala is an important mechanism to avert danger. Damage to this brain structure in monkeys leads them to behave in totally inappropriate ways such as approaching potential predators, like snakes, or other unfamiliar monkeys that they would have otherwise avoided.[110] The amygdala are a key part of our emotional brain circuitry, and how this part of the brain reacts depends on our genetics and past life experiences. Its activation is also altered in those suffering from mental health conditions such as depression, generalized or social anxiety and post-traumatic stress disorder.[111,112]

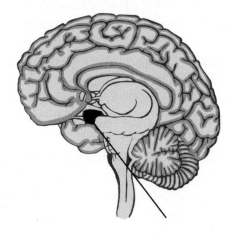

Amygdala
Emotional brain

The Emotional Brain: The amygdala are two almond-shaped structures that are activated when we experience emotions.

Not all our immediate emotional reactions are rational and acting on every single emotion we experience would be problematic, therefore the executive brain steps in and tries to rationalize them. This is called emotion regulation and, in many cases, the executive brain will regulate the emotional brain by applying a metaphorical brake. If you've ever been in a situation where you've felt angry, or any other strong emotion, but you've been able to calm yourself down, this is what is happening. However, emotion regulation does not just involve suppressing or eliminating emotions. For emotion regulation to work well, the executive brain and the emotional brain should have a bidirectional, collaborative relationship. Activation of the emotional brain provides the executive with valuable information: feeling strongly about something signals to the executive brain that there is something that we should be paying attention to. This is why, for example, emotional posts on social media or attention-grabbing headlines about smartphones, especially those employing alarmist tactics, tend to captivate us. After the emotional brain provides an alert to capture our attention, the executive brain combines this with information from other brain areas and past experiences in order to consider potential long-term outcomes and determine our actions.

Difficulties arise if we are in a state where the emotional brain is sending constant alerts, or the executive is having trouble processing them.[113] This could be due to being in a stressful situation beyond our control or having a mental health condition such as a constant state of anxiety. No matter the reason, if the executive has to continuously process emotional input and keep slamming that metaphorical brake on, it can be very draining. As we have learned, the executive brain is also responsible for our focus, motivation and willpower so, when excessive emotion regulation leads us into low power mode, these abilities begin to suffer. This will be more prominent in those who are either genetically predisposed to having less executive power or have conditions that lead to a degree of executive dysfunction. When the executive brain is exhausted, people are less able to exert self-control or be goal-oriented and the ability to regulate our emotions worsens. When depleted in this way, with a lack of processing power, people are more likely to take actions based on their emotional brain – they may find themselves getting upset, being irritable or losing their temper at something that is seemingly inconsequential.

If our internal resources are depleted, there is another strategy we can use to emotionally regulate – we can reach for external support. Much like you can lean on someone for support if you have difficulty walking, we can use other people's brains to help regulate our own. This is external regulation. If you've ever found that sharing your problems with someone was helpful, this is why. External emotion regulation is very powerful and explains how we feel comforted when we share our worries with others or receive a hug, and we seek to do so in return.

The emotional brain provides an important protective mechanism for survival, something that is especially important in young children. This is why the amygdala is one of the fastest developing brain structures, doubling in size in just the first year of life.[114] Since the function of the executive begins to develop later in childhood, the emotional brain in babies and young children primarily relies on external sources of emotion regulation. A parent's influence therefore has the ability to both decrease and increase activation in a child's emotional brain. For example, the comforting presence of a parent makes children better able to enter new situations or do things that are making them nervous. By soothing our children, we can reduce activation of their emotional brain, but there are equally situations where we need to increase it by alerting them to potential dangers. Children's tendency to form attachments to security blankets or cuddly toys also stems from this reliance on external forms of emotion regulation. As the executive brain begins to develop, it must initially rely on the amygdala – which had a head start in learning about the surrounding world – so its actions are therefore guided by a strong emotional component. Over time, the executive gets better at running the show but it is not until the transition into adolescence that more mature connections between the executive and the amygdala start to form. This coincides with the time that our children become significantly less reliant on us for understanding basic emotions but they may still need our external emotion regulation support with more complex ones.[115]

Powerful coping tools

One day, during my regular neurology clinic, something inadvertently caught my attention. Every single person in the waiting room was engrossed in their phones. You could come to the conclusion that this was a classic example of our unrelenting need to be on our phone every second, that our social fabric is unravelling while clutching our devices, but, instead, I realized that this behaviour made perfect sense in this particular setting.

Visiting a doctor can be an anxiety-provoking experience. Patients must share deeply personal information, often with a stranger, and the possible outcomes of the visit can be worrisome. The combination of this anxiety and the boredom, frustration and impatience that arise from waiting for a significant amount of time in a state of uncertainty can have a fatiguing effect on the executive brain, diminishing one's ability to emotionally self-regulate. Given these factors, it makes sense that people might instinctively reach for their phones as a coping mechanism.

Besides seeking help from other people, another key source of external emotion regulation can be found in technology. This phenomenon, known as 'digital emotion regulation', has been underscored in numerous studies which highlight our propensity to leverage technology for mood management and stress relief.[116] Digital emotion regulation could involve tapping into our digital support networks, such as texting a friend or engaging in online spaces where others share similar experiences, thereby fostering a sense of connection. Alternatively, it could mean playing a video game, watching an entertaining clip, or getting immersed in music, all of which offer a temporary escape to a different realm. Humour is also a commonly used emotion regulation technique and possibly the reason why watching and sharing funny videos is so popular. All of these actions provide a temporary distraction – this is a powerful emotion regulation technique that anyone dealing with a toddler having a tantrum has probably employed. The reason it works is because the passing of time has a significant effect on our emotions, which usually reach a peak in intensity and then reduce gradually. Time therefore provides a sense of perspective, which is why people may do or say things in the moment that they regret later. The activity in their emotional brain has reduced, giving their executive more capacity to objectively evaluate the situation.

Digital emotion regulation may also be one of the reasons that teenagers navigate towards technology. Adolescence is a time when our children become more independent and, as they progressively become less reliant on the adults in their lives, they also need to learn to manage their own emotional state. In addition to their chronotype moving later, as discussed in the previous chapter, resulting in the possibility of increased phone use at bedtime, digital emotion regulation may be a factor in how teenagers use their devices.

The potency of digital means for emotion regulation is illustrated by a 2011 military study, in which it was found to be useful for alleviating physical symptoms. The research found that soldiers playing a virtual reality video game while undergoing the very painful process of having their wounds cleaned and dressings changed experienced a considerable amount of pain reduction.[117] This is because pain does not just activate the sensory areas of the brain but the emotional ones too.[118] Knowing that you are about to have a painful procedure puts the brain on high alert and reduces your pain threshold so that you feel more pain. For the soldiers, playing the game provided a powerful and much-needed method of regulating the emotional components of the pain they were experiencing, helping them cope with a typically excruciating process which they had to go through on a daily basis.

Emotion regulation, undeniably, plays a pivotal role in our mental health and well-being. Yet, it's important to remember that this isn't a one-size-fits-all situation. What provides one individual with a sense of calm might not have the same effect for another and not every coping strategy is applicable or beneficial in every circumstance. So instead of providing a strict set of rules about when to employ digital emotion regulation and when it should be avoided, I would like to highlight some key concepts that can help you assess your individual situation.

Above all, it is important that you do not neglect strengthening your own internal machinery because digital emotion regulation should not be the only tool in your toolbox. While digital emotion regulation is not a new coping strategy – prior to the Phone Age, you could watch a film, listen to music, read a book – the widespread availability and convenience of smartphones have made them increasingly prominent as tools for managing emotions. Their easy access and breadth of content available means increasingly using our devices for digital emotion regulation, relying on them as a source of external support until they become our go-to action. Over-dependence on this type of emotion regulation can lead us to creating phone habits that we later find problematic – particularly as we have already learned that many of these habits can be triggered by emotional reminders. It is therefore important not to over-rely on digital strategies and to improve our own internal coping mechanisms.

What most people do not realize is that, if you want to reduce your emotional reactivity, it is far easier to start strengthening your own internal machinery outside periods of emotional stress than to wait for an emotionally-charged moment when your executive is inevitably under strain. To do that, the most constructive technique to deploy is mindfulness meditation – the process of non-judgementally paying attention to the present moment. Practising mindfulness is to consciously focus on immediate experiences whether it is your breath, body posture, sensations, thoughts or emotions. Every time your mind wanders, you should gently redirect your attention back to the present (without feeling bad about it). Over time your brain will get better at this skill meaning you can apply similar techniques during times where emotions are heightened. Techniques from this book such as the Advanced Five-minute Rule: Surf the Urge (Building Block 1) feed into this strategy as they make you non-judgementally assess your emotions and sit with the discomfort of not reaching for your phone in a time-limited manner. In doing so, you make more intentional actions and hone your own internal machinery in the process.

Neuroscience supports the benefits of meditation as an effective technique for stress reduction. A lot of anxiety and worry stems from our mind being overly focused on the future and an influential scientific study showed that we are happiest when our mind is oriented in the present moment.[119] While we can use the power of anticipation (as described in Building Block 7) to look forward to rewards, we are less happy when we are thinking about negative, or even neutral, topics in the future than when we remain focused on the present. Practising mindfulness can help reorient us to the present moment. Moreover, those who meditate regularly show reduced activation in the amygdala and changes to the structure of the executive.[120] It is the equivalent of strength training for your brain: not only does mindfulness meditation make the executive stronger and better able to handle more pressure but, at the same time, the emotional brain becomes less hyperreactive. This approach further eases the strain on our executive function, thereby boosting its available resources. Consequently, it facilitates better attention span and amplifies willpower levels, both of which significantly influence our habits in a positive way.

Another potent method for reinforcing our internal emotion regulation framework is movement. Physical activity can serve as a constructive coping mechanism to counter stress, offering the same form of distraction that we seek when we reach for our smartphones, but with the added advantage of triggering the release of endorphins – our body's natural mood lifters. Beyond this immediate impact, regular exercise encourages better sleep which, as I have repeatedly emphasized throughout this book, recharges our executive capabilities. Additionally, physical activity prompts the release of nourishing substances, such as brain-derived neurotrophic factor (BDNF) whose job is to nourish our neurons.[121] Enhancing these benefits by spending time in nature amplifies the positive effects of exercise; an hour-long walk in natural surroundings has been shown to decrease activation of the amygdala.[122]

If you don't have the time to take an hour-long walk, you shouldn't view smaller actions or interventions as not worthwhile because they still have a sizeable impact. The research looking at the effect of walking in nature relied on costly scanning procedures, hence the need for substantial interventions – like the hour-long walk – to justify the investment and yield results. But it's not an all-or-nothing scenario – as we saw in the 80/20 Rule (Building Block 7) going from 'nothing' to 'something' has the biggest effect. Incorporating meditation and movement even in small amounts into our daily routine increases the power of our executive. This increase in power can then, during emotionally-charged moments, provide a pause that we can utilize to change our response. The ability to pause when we encounter an emotional reminder helps break apart the first two puzzle pieces of any problematic habit. Thereafter, using these pauses to insert small periods of mindfulness and exercise starts to build supportive Really Small Habits. This not only helps reduce your emotionally-driven phone checks but also encodes alternative pathways, which will eventually substitute any contradictory habits (Building Blocks 8 and 9).

Once you've honed your emotional regulation mechanisms during moments of calm, you can then reduce your dependence on digital emotion regulation during high-stress periods by incorporating some non-digital techniques. When the amygdala is triggered, it signals to our body to release fight-or-flight hormones, activating our sympathetic nervous system, which in turn increases our heart and breathing rates. During intense stress, use the reflective pause provided by your strengthened executive to practise breathing exercises – examples are deep diaphragmatic breathing, box breathing, 4-7-8 breathing, five-finger breathing or a physiological sigh. The specific breathing exercise matters less than the act of modulating the breath itself. Adjusting your breathing essentially activates the parasympathetic nervous system, which then counteracts the sympathetic response, and provides feedback to the emotional brain to dial down its intensified activity. Incorporating journalling during this reflective pause can also aid in rationally processing these emotions. Ensure you use pre-commitment (Building Block 4) to plan what techniques you will implement ahead of time. While digital emotion regulation has its place, these techniques empower you by fortifying your own innate emotional processing capabilities.

At times, it's possible to engage in both digital and non-digital emotion regulation simultaneously – for instance, listening to a podcast while walking – or sequentially, such as doing some breathing exercises followed by some digital distraction. Pairing these activities can enhance habit formation, as demonstrated in Temptation Bundling (Building Block 10). While starting any new habit can feel challenging, the very process of habit formation also strengthens our executive and gives us the all-round benefits of better executive function.

While making sure that you continue to strengthen your internal machinery, you should also consider the types of situations in which you employ digital emotion regulation. In general, using digital devices may be a good coping strategy during a set of circumstances that are out of our control and where the mere passing of time will help. However, as it is a passive technique, it may not be a useful tool if a situation requires active input. Consider the following question – would time passing resolve the situation? If the answer is 'yes' then digital emotion regulation is appropriate but when it is a 'no', you probably need to tackle the problem in a different way. For example, many of us go through intermittent periods of feeling unwell during which we are able to do little else besides stay in bed. In those circumstances, which are out of our control, using technology can be a good distraction from our symptoms. Many of my patients will use both digital and non-digital activities to take their minds off their hospital stay. Likewise, all of us experience fleeting periods of stress, sadness or anxiety, which will typically resolve on their own. It is important to note that the stress or discomfort that you are managing must be mild as more severe difficulties will need more active input including professional help. However, the use of technology can become detrimental if our recovery from a certain illness, physical or mental, relies on taking action but our digital habits contradict this by promoting inaction. In those cases, scrolling on our phone is detrimental if it either prevents us seeking help or if it is at the expense of engaging with elements of recovery.

Rather than a coping mechanism, it becomes a form of avoidance. Recognizing the power of our technological coping tools and becoming more self-aware will allow us to be more discerning in when and how we use them.

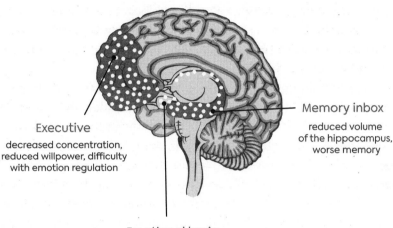

Executive
decreased concentration, reduced willpower, difficulty with emotion regulation

Memory inbox
reduced volume of the hippocampus, worse memory

Emotional brain
enlargement of the amygdala potentially leading to increased emotional reactivity

Effects of Stress on the Brain: Chronic stress leads to a dysregulation in cortisol affecting these three key areas of the brain which have high cortisol receptors.

So, it turns out that being in a doctor's waiting room is a great example of when digital emotion regulation is helpful. As I smiled and greeted my next patient, I reflected on how easily we make unfounded assumptions about people's behaviour. Finding myself in the same situation a few months later, in a similar waiting room, waiting to undergo an anxiety-inducing procedure, my phone buzzed. It was a message from a good friend. I immediately responded, telling her about my worries and at the same time accessing an external source of support that I needed. But to the outside world, I looked just like those people in the waiting room, just another person 'glued' to their phone.

Digital stress

The multifunctionality of our devices and the complex role that they play in our lives mean that they can simultaneously be a coping mechanism but also a source of stress. It is not the technology itself that causes anxiety and pressure, but a combination of societal demands, the content that it enables us to access and the habits that we have developed. It is also important to note that stress is not always detrimental. A small to moderate amount of stress can be a very effective motivator and even enhance our performance. Problematic stress occurs when the demands placed on our brain exceed the available resources our brain has to manage them. At excessive levels, the emotional toll of stress can put our brain into low power mode. As we have discussed previously, this can be remedied by engaging in replenishing activities or getting enough sleep, but if problematic stress is chronic and recovery is inadequate, there is a risk of becoming burnt out. While many of the symptoms of burnout can reflect those of low power mode, it is in fact a deeper state of exhaustion, which takes longer to recover from.

While technology has allowed us to stay connected with our loved ones and access our external support network, it can also create a situation where taking a restful break away from work is difficult. As explored in Chapter 9, high demands at work and, for example, email-checking habits, fuelled by societal expectations of constant productivity, have blurred the boundaries between work and rest. We feel we need to fill every spare moment we have and to be online the whole time. Our portable devices enable this while leaving less opportunity for us to replenish our executive brain. Constantly checking emails may not be a problem for everyone so we shouldn't overgeneralize as the impact of this habit will vary depending on the workplace, the type of work and the individual brain involved. Some will find their employment energizing and fulfilling to a degree that means they can work long hours without becoming depleted, but many will at some point experience digital stress arising from work, which can take an emotional toll.

Outside of work, we may reach for our phone for entertainment or a degree of escapism but, depending on our habits, this may not have the effect that we intend. Our emotional brain is inherently wired to be captivated by emotive information, with fear being particularly effective at eliciting strong activation. This is a survival mechanism. Our digital habits often mean that we are constantly refreshing either the news or social media, so that we are regularly drip-fed emotionally charged information, prime examples being upsetting events occurring around the world or argumentative comments on social media. Reading about a threatening global event such as a pandemic, a war or a natural disaster can make us feel helpless and regulating the high level of emotional activation it elicits in our amygdala is draining in itself. Yet our response doesn't tend to make things any better. As we've learned, uncertainty imposes a significant burden on the executive. One coping mechanism we employ to regain control is to seek more information about the uncertain event. We comb through various sources in the hope of unearthing a piece of information that will help make sense of the situation. This compulsive consumption of negative news has been referred to as doomscrolling and does not just occur with global events but with personal ones too, such as a health scare. While using this method of information gathering to combat uncertainty may trick our brain into feeling slightly more prepared in the short term, beyond a certain point there are diminishing returns and information is no longer helpful. In fact, it can have adverse consequences for both our mental and physical health over the long term.

The general mental preoccupation with what is happening in the digital world is referred to as 'online vigilance'. This constant preoccupation with online content and communication can have a negative effect whether we are engrossed with our work emails, likes and followers on social media or threatening events on the news. Devoting a significant portion of our limited cognitive resources to constantly patrolling the online world – whether by direct interaction or a constant mental preoccupation – hinders our brain's capacity to remain in the present moment. This also depletes the executive reserves we need to handle other challenges, making us more susceptible to feeling overwhelmed.

If our digital habits mean that we have a near-constant exposure to stressful content, we add another draining effect to our brain. In a healthy individual, levels of cortisol are usually highest in the morning and decrease at night. In acutely stressful situations, the brain provides instructions for the adrenal glands to temporarily release the same hormone as part of our fight-or-flight response, a short-term mechanism to increase our alertness. However, at times of chronic stress, which includes prolonged high levels of cortisol, this morning-evening variation of cortisol is lost. Instead of being manufactured 'on-demand', what we see is a steady supply being released throughout the day. This desynchronized, non-physiological pattern has a considerable impact on both our mental and physical health.[123]

Three key brain areas – the executive, the memory inbox (hippocampus) and the emotional brain (amygdala) – have an abundance of cortisol receptors and are therefore significantly influenced by changes in cortisol levels. Increased cortisol can notably impact our executive function, often manifesting as decreased concentration during periods of prolonged stress. As our executive power reduces, the likelihood of entering low power mode increases. Consequently, we are more prone to acting based on the habits stored in our autopilot system and making decisions based on short-term rewards. Chronic stress can affect the hippocampus, causing a reduction in volume and subsequent impairments in memory. Being under constant stress has also been shown to lead to an enlargement of the amygdala, potentially altering our emotional responses.[124,125] Coupled with reduced emotion regulation from the executive, this means that, at times of chronic stress, we will find ourselves constantly on edge or getting angry and upset by situations we would ordinarily be able to cope with more easily. At the same time, cortisol remaining high at night disrupts our sleep and its vital restorative functions, and thereby indirectly affects the whole of our brain. Chronic stress, whether digital or non-digital, is fatiguing and starts a vicious cycle that makes us more prone to develop problematic digital habits.

Although having good digital habits can help, a certain proportion of digital stress is something we cannot avoid, whether it comes to workplace demands or threatening events such as pandemics and wars. However, evaluating whether you suffer from variations of online vigilance, such as a preoccupation with your email inbox or gaining likes or followers on social media, can be helpful when trying to determine and therefore tackle the cause of any anxiety or chronic stress. The 'always on' mindset could be one of the reasons your technology is having a negative impact on you.

Screen time is a symptom

Usually, an alarmist headline regarding phone use will quote scientific studies to support its point. The typical quoted study will compare a group of people with a mental health issue, such as depression or anxiety, to a group without that condition, and report that the former spend more time on their phones. While it is true that some studies do support this, we are being presented with a skewed picture. Sensational or worry-inducing headlines activate our emotional brain and are more likely to capture our attention, and for that same reason, negative findings tend to receive more coverage, even when it comes to publication in scientific journals. This underplays the studies that don't find a link between poor mental health and increased phone use – of which there are many – and which go unreported as they are deemed uninteresting. There is nothing to report, as such. But, even if all the studies did agree, does showing that people who are anxious and depressed spend more time on their phones provide proof that their phone has caused a mental health problem?

The types of comparative studies described above are known as cross-sectional studies. They have the advantage of being quick to conduct and can generate hypotheses for further investigation, but they cannot provide conclusive evidence. In these studies, we can observe an association between two variables, such as phone use and mental health problems, but we cannot determine which variable is the cause and which is the effect. For example, is excessive phone use causing mental health issues, or are mental health issues leading to increased phone use? Or is there a third factor, such as societal pressures depleting our executive resources and putting us in low power mode, that is causing both increased phone use and mental health issues? Or even, could being exposed to alarmist newspaper headlines be contributing to mental health problems?

It may seem like I am being picky, but this is not the case. The phrase 'association, not causation' is frequently repeated in both medical and scientific communities, to highlight this exact limitation. It serves as a fundamental principle that we should not hastily draw conclusions without sufficient evidence. An example that epitomizes the concept of association without causation is the relationship between increased ice cream sales and an increased number of shark attacks. Clearly, eating ice cream will not cause you to be attacked by a shark. This association is not causative; the underlying factor of both phenomena is hot weather. During warm summer months, people are more inclined to enjoy an ice cream while, simultaneously, the warm weather also attracts individuals to swim in open water, increasing the likelihood of encountering sharks.

Similarly, comparing groups of individuals who are suffering from mental health issues to those who are not will reveal many differences between them. But it is impossible to tell what is cause, what is effect and whether there is an underlying third factor influencing both of those variables. To begin establishing this cause-and-effect relationship, it is necessary to track a group of individuals without pre-existing mental health conditions over an extended period of time and observe whether increased phone use precedes the development of mental health conditions. These are called longitudinal studies, which are much more difficult to conduct than cross-sectional studies due to the length of time and expense that they require. Whereas cross-sectional studies can be done fairly quickly, providing a quick snapshot between two groups at any point in time, longitudinal studies involve the passing of time. This is why there are many more cross-sectional studies than longitudinal studies out there but, no matter how many imperfect cross-sectional studies we do, it is impossible to create perfect data. We are actually more likely to be led down the wrong path and reach the wrong conclusion.

The lengthiest longitudinal study to date monitored 500 participants for eight years to see what effect social media had on their mental health. Their ages, thirteen to twenty, made them specifically vulnerable as many mental health conditions become apparent in adolescence or early adulthood. Rather than comparing two completely separate groups, one of which had developed mental illness, and trying to work backwards, they worked forwards. This method meant asking participants to report their social media use while they monitored symptoms of depression and anxiety through questionnaires. At age thirteen, participants spent thirty-one to sixty minutes on social media per day and this increased fairly steadily so that, by young adulthood, the average was more than two hours per day – a significant amount. Throughout the study, participants' social media use would naturally ebb and flow but there were no corresponding changes to their mental health as a result. In other words, spending more time than average on social media did not lead to a subsequent increase in depression or anxiety scores and reducing social media use did not yield a benefit. The researchers of the eight-year study concluded that they found 'no evidence that social media use is "destroying a generation"' as we are so often told.[126]

It is worth noting that if the researchers of the eight-year study had used the snapshot approach of cross-sectional studies to compare two groups, they would have found the same association between social media use and depression/anxiety scores that the sensationalist newspaper headlines frequently flaunt. However, without the insightful dimension of time, this approach could have led them to an erroneous conclusion. The longitudinal approach allowed for a more accurate analysis of the relationship between social media use and mental health. That being said, this study is not perfect and has its own limitations. For example, participants were asked to report their own screen time and symptoms rather than undertaking objective assessments, something which is often less accurate but is a good example of the difficulties that studying human beings entails.

So, when you read alarmist headlines, rather than jumping to conclusions that phones worsen our mental health, it is important to keep in mind that phone use may well be the consequence rather than the cause. Given what we have learned about digital emotion regulation, people with anxiety and depression could be reaching for their phone as a coping mechanism. It is more likely to be a symptom of their struggles. Patients with severe depression that I encounter in clinic often find it difficult to get out of bed but proclaiming that it is the comfiness of their bed that has led to their depression would be absurd. The difference between the bed and their smartphones is the collective feeling of unease that we feel about our devices, which clouds our judgement. It makes us more likely to accept an outcome that fits with our strong feelings and makes us more likely to rely on low-quality scientific evidence.

But the picture is probably even more complex and nuanced and difficult to untangle. This is because often, during times of difficulty, we engage in behaviours that inadvertently worsen our situation. Just like if we were to stay in bed all day, spending a significant amount of time scrolling can prevent us from engaging in activities that have a positive mental health impact, such as physical activities, pursuing hobbies and socializing with friends. And while not being addictive, there are several features of phones and apps that make them habit forming so, used incorrectly, phone use may be having an indirect effect on mental health. Separating these factors is a challenging task that will require further detailed research. Meanwhile, the next time you encounter a sensationalized headline about phone use, bear in mind that these are often misleading, often failing to mention the underlying scientific study's limitations.

The oversimplification of mental health

There has been an undeniable increase in the diagnosis of mental health conditions in recent years. It is not uncommon to be confronted with a graph that depicts this growth, along with a vertical line placed around 2007–10 at the point these diagnoses start to rise, showing when smartphones became commonplace. But this is a superficial way of looking at things and, as illustrated by deconstructing the studies above, putting the rise in mental health difficulties down to a single factor is too simplistic. During that time there has been a global financial crisis, wars, natural disasters, a changing political landscape as well as growing concerns about climate change, to name just a few distressing events that can affect mental health.[127]

Greater awareness and understanding of mental health issues, as well as the reduction of stigma around having conditions such as depression and anxiety, has meant that people are more likely to seek help and open up about their problems, something I have personally witnessed in my own medical practice. This has unsurprisingly resulted in an increased number of diagnoses being reported in our statistics. And we shouldn't forget that smartphones are used to deliver mental health interventions and have increased accessibility to mental health services, not to mention that social media and online forums have provided environments for informed discussion and shared experience. Many mental health helplines reach a greater proportion of people by providing an option to communicate via text which is more appealing to young adults, neurodiverse individuals or anyone who finds it difficult to open up about their problems.

Considering smartphones and the access they provide to the online world to be the main factor contributing to worsening of our mental health is undoubtedly a vast oversimplification of how mental health problems work. They are multi-factorial and depend on a combination of genetic vulnerabilities and environmental life stressors. This is certainly the case in the patients that I see. Having a family history of mental health illness indicates a genetic susceptibility, but my patients' stories also reveal very difficult life circumstances, and the list is long. They may have experienced social and economic inequalities, been the subject of racism or other discrimination, be living in poverty or in unsafe areas with increased pollution and less access to green spaces and healthy food. They may not have an adequate support network, family relationships may be strained or non-existent, they may have experienced domestic violence or early childhood trauma. Even if it turns out that smartphones are having a detrimental effect on our mental health, this is likely to be very small and to be vastly outweighed by more significant factors. This point is illustrated in a 2019 study by the University of Oxford. Dr Amy Orben, together with Professor Andrew Przybylski, analysed over 355,000 adolescents from three large databases in the UK and the US to look at the usage of digital technology (divided into categories like social media, computers, Internet use, among others) on mental health. To ensure the reliability and validity of their findings, the researchers employed a pre-registered analysis. This method prevents excessive calculations on a dataset, eventually choosing the one that fits a predetermined narrative – an issue that has potentially affected the results of many research studies on smartphone use. The result? Technology use accounts for a mere 0.4 per cent variation in well-being.[128] It's critical to note that this is still a correlation, not causation – it's possible that screen time could be symptomatic rather than causal. But the microscopic scale of this correlation, especially when contrasted with the overreaction in the media, is astounding. Binge-drinking, smoking cigarettes, using marijuana, being bullied, or getting into fights all had associations several magnitudes greater than technology did. Even the correlation between wearing glasses and well-being was more significant than that with technology use.

Expanding on this and focusing specifically on social media, Professor Andrew Przybylski, who studies human behaviour and technology at the Oxford Internet Institute, collaborated with Professor Matti Vuorre to assess well-being data from nearly a million Facebook users. This extensive research – arguably the most comprehensive on social media and mental health – spanned twelve years (from 2008 to 2019), covered seventy-two countries, and involved 946,798 participants.[129] Yet, it revealed no concrete evidence linking social media adoption to psychological harm. This lack of harm was consistent across different age groups and different countries. Given its comprehensive scope, this research stands as the most conclusive evidence we currently possess that social media is not responsible for any worsening in mental health that we are seeing on a population level, effectively counteracting the media narratives that claim social media is wreaking havoc on our mental well-being.

This is important because, while the association between technology use and well-being was shown by these studies to be very small and even non-existent, the disproportionate negative coverage related to phone use and mental health issues can still have an impact on us. Our expectations have such a powerful effect on our brain that they have the potential to become reality. We commonly see this in medicine through the placebo effect. The impact on pain relief of being injected with water is contingent upon the participants' beliefs and expectations. If subjects believe the injection to be a potent painkiller, they will experience significant reduction in their pain. The participants are not lying about their experience. The mere expectation that the injection will relieve pain causes the brain to react in such a way that it produces its own natural chemicals leading to a deactivation of pain processing areas.[130] The subjects are, in fact, experiencing true pain relief.

On the other end of the spectrum from the placebo effect lies the 'nocebo effect'. Statins are a drug used to prevent the build up of high cholesterol in our blood vessels and hence reduce dangerous conditions such as heart attacks and strokes. Much like our phones, there has been a lot of negative press coverage regarding their side-effects. To investigate this, a study published in the *New England Journal of Medicine* recruited participants who had discontinued their statin treatment due to severe side-effects. Over the course of the trial, participants alternated between taking a statin and an inactive sugar pill, each for a one-month period. Strikingly, even when on the sugar pill, participants reported experiencing nearly all (90 per cent) of the same severe side-effects that had initially led them to stop their statin treatment.[131] It's not that statins do not have side-effects; they certainly do. But a strong belief that a drug will have harmful side-effects makes them more likely to be recreated by our brain even in the absence of that drug through the 'nocebo effect'. This insight meant that a considerable proportion of patients who had stopped their statins following worries about side-effects were happy to restart them after taking part in that trial.

In a similar manner, scrolling on our phone while constantly fearing that it will have a toxic impact on us will activate our amygdala and eventually become a self-fulfilling prophecy. We do not realize that our anxiety and fears are not a result of our phone use itself, but because we are consumed with worry and guilt. These feelings of shame at the inability to control how we use our so-called 'toxic' phone have the potential to directly impact our mental health. This idea was effectively highlighted in a study of 245 participants where mental health difficulties were associated with scores on questionnaires related to problematic phone use. Interestingly, when objective measures of actual phone use were taken into account, this relationship disappeared. The study revealed that participants who expressed more concern about potential 'problematic use' were more likely to have mental health difficulties, even though they didn't actually spend more time on their phones. The authors of the study even concluded that 'addressing people's appraisals including worries about their technology usage is likely to have greater mental health benefits than reducing their overall smartphone use'.[132]

There are a lot of contributing factors that outweigh our relationship with technology, and I have become increasingly uneasy about the constant narrative that places a disproportion-ate amount of blame for mental health issues on smartphone use. This type of narrative can be used in a way that covertly blames individuals. It is as if to say, 'if you didn't use your phone so much then you wouldn't have a problem'. This focus on phones also distracts us from addressing the more significant issues that may be affecting our mental health. This is why, rather than fuel the same scaremongering rhetoric, I wrote this book to restore some balance and to make us consider our relationship with our phones more critically.

Mental health practical

Emotion Regulation Rules

- Use a combination of internal and external emotion regulation techniques.

- Ensure you strengthen your internal machinery through mindfulness and exercise.

- Digital emotion regulation can be used as part of your toolkit.

- Avoid passive digital emotion regulation when active involvement is needed.

Emotion regulation techniques

USE PLAN B TO HONE YOUR INTERNAL EMOTION REGULATION

PLAN A	BRAIN BENEFIT	PLAN B (FOR WHEN PLAN A IS OVERWHELMING)
Mindfulness meditation	Reduces activation of the emotional brain and, long term, boosts the executive brain	Breathing exercises*
Going for a run	Releases feel-good chemicals including BDNF which nourishes our neurons	Going for a walk
Spending time in nature	Regulates the emotional brain	Tending to some houseplants
Writing in a journal	Processes emotions	Mentally identifying and labelling your emotions
Practising yoga	Promotes mind-body connection, reduces stress	Single yoga pose or some simple stretches
Doing a complex puzzle	Engages the executive brain, promotes problem solving	Doing a Wordle or Sudoku

MANAGE DIGITAL EMOTION REGULATION

Even when it comes to digital emotion regulation, having a variety of techniques is important to reduce problematic habits surrounding a single app.

* Examples are deep diaphragmatic breathing, box breathing, 4-7-8 breathing, five-finger breathing or physiological sigh. Descriptions or demonstrations of each breathing technique are readily available through online sources.

If you constantly reach for the same app as a result of emotional reminders, then try the following emotion regulation techniques:

- Create a 'smile file' – this can be an album on your phone with nice pictures, videos or quotes that you can enjoy revisiting. Screenshot any nice messages you receive and keep them there for a boost.

- Message a friend.

- Use the notes app to write down your feelings.

- Play a simple game.

- Listen to music or an audiobook.

- Use a mood-tracking app to enter how you are feeling.

- Do a meditation using an app.

Setting boundaries

- If you can relate to the section on digital stress and feel like the constant exposure to potentially threatening events in the news, argumentative comments on social media or the stress of your inbox is affecting your mental health and your sleep, then Insert Hurdles (Building Block 3) or Leverage Location (Building Block 5) to create boundaries.

- When communicating your boundaries, use the right language to protect them. A study showed that you are more likely to stick with these changes if you switch to saying 'I don't' instead of saying 'I can't'.[133] Rather than 'I can't check my email outside work', try 'I don't check my email when I am spending time with my family.'

★ It is also important to make sure that you are getting adequate rest and recharging your battery. If you are in low power mode, you will have fewer resources available for emotion regulation.

*

Please note that there is a difference between looking after your mental health by reading self-help books and suffering from a mental illness. Mental illnesses, just like physical illnesses, will need professional help and so please reach out to a medical professional.

12

Social Media

My formative years were free from the ubiquitous presence of smartphones; they were instead defined by the omnipresence of water. Growing up in Greece, where I spent summers by the sea, I learned from my mother the significance of respecting the water and, from an early age, great importance was placed on acquiring the skills to navigate it safely while understanding my physical limits. Above all, my mother's teachings were always guided by knowledge rather than fear.

I often think of my mother's advice when considering the digital environments and challenges that we have to navigate today. I did not grow up with social media but similar to the vast expanse of water that once enveloped me, it has become an inextricable part of our lives, an undeniable force that surrounds us and which cannot simply be wished away. As a neurologist and neuroscientist, I find myself captivated by how this expansive digital landscape impacts the functioning of our brains. And, as the mother of two young daughters, my own upbringing and my mother's teachings continue to echo in my mind. *Let us not fear the water but instead learn to swim.*

Over the last couple of decades, concern about social media use and its impact on us has emerged as a specific area of focus – and fear. These rising anxieties are the reason I devoted an entire chapter to the topic, and it has been the chapter that was most challenging to write. Social media platforms have revolutionized the very fabric of our connections, communication and information consumption. They have granted us unprecedented opportunities for sharing, interaction and discovery. Yet, within the digital realm, we also encounter distinct challenges that shape our cognitive processes.

Not only is there a lot to cover, but the debate about the impact social media has is so heated that it tends to come with the pressure to take a side. To pass definitive judgement and label social media 'good' or 'bad' once and for all. Staying off it at all costs is often encouraged as some sort of personal victory. But what if you don't want to? When asked their opinion, the majority of people feel that the positives of social media outweigh any negatives.[2] I personally use social media to catch up with my friends, to share my professional experience, and I found it a useful tool to help me navigate the earliest stages of parenthood, all clear positive experiences of social media. Nevertheless, social media has pitfalls. This includes grappling with issues related to the content we consume, its influence on our self-perception, and the potential development of problematic habits associated with both content consumption and posting. The advantages and drawbacks of social media are in fact more nuanced than what media and popular opinion might lead you to believe.

Moreover, having such strong negative opinions actually prevents us from fully benefiting from the positive aspects of social media. Saying we should give up social media is both disempowering and, ultimately, unrealistic for many. I propose a more balanced approach where we acknowledge there are good and bad elements to social media, and through giving you some valuable insights into the complex relationship between social media and our brains, I hope to empower you to build healthy digital habits that leverage its potential benefits while protecting you from any negatives. By combining scientific knowledge with practical strategies, we can equip ourselves with the skills needed to navigate the ever-changing currents of social media. The opposite of using technology badly is not to give it up; it is to learn how to use it wisely.

Content

Our brain is not an unbiased observer of the world. It selectively filters and processes information based on its relevance and significance to us. When I became pregnant, there were suddenly more parents walking with their children in pushchairs. Why had I not noticed them before? I am sure you have experienced similar examples in your own life. The most common is getting a new car – cars of the same make and model suddenly stand out more than they did before.

While you go about your day-to-day life your brain is undertaking an enormous amount of processing. Thousands of pieces of information reach our brain every minute: our eyes can see hundreds of objects in our surroundings; our ears can listen to a multitude of sounds; and, internally, our brain is receiving information about our different bodily functions including our heart rate, digestion and temperature. But, we are not aware of all of this, as most information is filtered at the subconscious level, and we only pay attention to a select few elements.

The reticular activating system is a highly complex network of connections that runs throughout our brain, from the base of our brainstem to the executive behind our forehead. One of its many functions is to regulate our attention, ensuring that we focus on what is important.[134] This system works by shining a spotlight on relevant information and filtering out everything else. For instance, even in a noisy and crowded room, hearing your name will make you perk up, a phenomenon known as 'the cocktail party effect'. Directing this attention spotlight onto a particular conversation means your brain will turn up the volume on this item and reduce irrelevant background noise.

While we can direct this spotlight consciously, our internal state, thoughts and emotions can also influence the attention spotlight subconsciously and direct its focus to particular aspects of our surroundings. Undergoing a life-changing experience, such as having a baby, will occupy a significant portion of our cognitive resources, directing this spotlight so that we are more likely to notice other new parents. Any activation of our emotional brain also sends a powerful alert to direct the spotlight. This is what makes us instinctively drawn to emotionally charged content on social media, and why those types of posts garner the most reactions.

The digital world, like our physical environment, produces an overwhelming amount of information. The average social media consumer follows too many people to be able to see all the content on their feed every single day. To address this issue, algorithmic feeds – as discussed in Chapter 7 – are now deployed by social media apps. The algorithm that determines what we see ranks content based on importance, which respects our limited attention by showing us 'the best' or the most relevant content rather than the 'most recent'. However, this ranking also creates a key issue that we need to be mindful of. Unlike the relative constancy of the physical world, the algorithmic nature of these feeds means they are able to change in accordance with our spotlight. An algorithm will use information about content that you have previously interacted with – such as liking, commenting or sharing – to determine the future content it shows you, meaning your attention spotlight plays a crucial role in determining the content you focus on. This information is used by algorithms to prioritize similar content to continue to capture your attention, resulting in a powerful amplification effect.

These algorithms are simplistic compared to the complexity of the human brain – they will show you what you *want* to see but this is not necessarily the same as what you *need* to see. This amplification effect is indiscriminate. It can fuel a passion, but it can also exacerbate a vulnerability. Our attention spotlight may be activated because we are interested in something but also because we are hurting. Being bombarded with parenting content may be useful for an expecting parent but not for someone who has suffered a miscarriage or is going through fertility issues.

As we learned in the previous chapter, having a mental health condition or a past history of trauma can significantly alter the activation of the emotional brain, and this in turn will change how this spotlight functions even more drastically. As the emotional brain is linked to our memory inbox, it is able to flag particular memories as important, making them more likely to be saved. We are more likely to remember memories laden with emotional content and, due to the amygdala's heightened response to negative stimuli, negative experiences are often encoded more deeply than positive ones.[135] This inherent negative bias is common to all of us, but those living with depression may experience even greater alterations in how their emotional brain functions, resulting in an even stronger inclination to remember negative stimuli, particularly if these are personally relevant.[136,137] This means that their memory will have an even more prominent negative bias on a subconscious level. Similarly, anxiety can lead to biases, but in this case, it is towards explicitly threatening information.[138] Digital algorithms, oblivious to an individual's personal biases and mental health state, might make use of both conscious 'likes' and comments as well as unconscious interactions – such as the duration one lingers over a piece of content without tapping the 'like' button, the scrolling speed, or the time spent on a specific post or page – to determine future content. Consequently, these algorithms could inadvertently present a vulnerable mind with an increasingly skewed perception of the digital world.

How our brain works therefore determines how we react to the content that is presented to us. Let's take a particularly pertinent example that is commonly discussed regarding social media: body image. Vulnerabilities surrounding body image are likely to be pre-existing, and the combination of our attention spotlight and how algorithms work means that these insecurities can be amplified. Someone with a distorted perception of their body image is more likely to shine their attention spotlight on to and engage with idealized images of 'perfect bodies' only to be presented with similar images at an increasing frequency. This creates a one-sided and misleading version of their digital world, which then continues to affect their spotlight.

When it comes to body image, research also shows that, rather than time spent on the platform, it is the type of content we are exposed to that has the greatest influence on us. There is a stark difference between having a social media feed inundated with images of 'ideal' or 'perfect' bodies versus one filled with funny video clips. Viewing content that glorifies weight loss to an unhealthy degree has been shown to be particularly harmful. Hashtags like #thinspiration are quite obviously linked to this type of content, but other hashtags such as #fitspiration and #strongnotskinny, which were initially started to counter the super-skinny narrative, have been found to encourage unhealthy behaviours such as restricted eating and excessive exercise, as they still promote an idealized body image, albeit just a muscular and toned one.[139] Research also shows that editing images and relying on beauty filters can exacerbate body image issues not only for those viewing these images but also for creators.[140] For a creator, receiving rewarding likes and comments after posting a filtered image can reinforce the false notion that editing is necessary, the brain's neuroadaptation system setting a new baseline level where editing is the norm so that one's unedited self appears worse by comparison.

While social media and algorithms bear the brunt of the blame for many of our issues, it takes more than just the content on our little devices to shape and influence our brain. Again, using physical appearance as an example, idealized beauty and body standards were prevalent prior to social media and have long been propagated by both traditional media and the people around us. So, when we see these unrealistic standards and images all around us, what we are actually seeing is social media reflecting our society back at us. It should be no surprise that people who grew up with unrealistically perfect and digitally altered images on the covers of magazines, and who internalized those beauty standards, have gone on to post content that mimics those unattainable images.

I've described a classic example of the pernicious aspect of social media but there is a 'good' side too where new types of content have been instrumental in helping us navigate away from narrow body ideals. Unconstrained by a limited group of traditional media gatekeepers, many people on social media have started to combat unattainable ideals and promote body acceptance and diversity. Research shows that these types of posts and images have a positive effect on those who view them and increase body satisfaction. Content directly comparing edited and unedited images to demonstrate how a picture can be altered via digital tools, filters or strategic positioning helps increase digital literacy and awareness, something studies show has a positive impact.[141,142] The grass-roots nature of this type of content has played a critical role in changing the dialogue around body ideals. You could argue it has even been instrumental in traditional media expanding the diversity of images found on magazine covers.

While body image is just an example and one of many issues that is commonly discussed when it comes to social media, there is a common theme. Social media content will reflect, and algorithms will amplify, deep-seated personal or societal issues while simultaneously representing the activism that aims to create change around these same issues for the better. It is usually the negative side of social media which receives focus, and which can be somewhat exaggerated compared to its positive aspects.

The way that social media algorithms alter our digital world means that they have been accused of amplifying and reinforcing polarized viewpoints, ostensibly creating 'echo chambers'. While this phenomenon does occur, its scale and the number of individuals it truly impacts are often overstated. This is not just down to the personalization of content that algorithms provide but our own ingrained confirmation bias which means that we self-select sources that are in agreement, and hence echo, our existing viewpoints. However, we live in an era of digital abundance teeming with a diverse array of sources. This was shown in a study of 2,000 UK adults exploring their consumption of political information, which revealed that, rather than being stuck in an echo chamber, most people have a surprisingly varied digital information diet.[143] Despite its broad reach, social media is often regarded as a less reliable news source, with many users acutely aware of the misinformation frequently propagated on these platforms. As a result, people tend to diversify their media consumption habits, relying on an array of sources and fact-checking information outside of social media – this is something that I strongly recommend you do also. Only 8 per cent of the adults surveyed were thought to have a less diverse information diet, making them potentially susceptible to the echo chamber effect. In other words, while social media can influence our views, it's just one piece of a much larger, diverse media puzzle that shapes our understanding of the world.

Our perception of social media as predominantly negative largely stems from how our brain functions. As we discussed earlier, the inherent negative memory bias means we're more inclined to remember and dwell on adverse experiences. This is essentially a safety feature, designed to help us learn from and avoid repeating such events. Take, for example, our experiences while driving. We'll distinctly remember a 'near miss', yet we don't form equally strong memories – or any at all – for the countless times we drove the same route without incident. This bias carries over to our interactions on social media. Social media platforms often get a reputation as breeding grounds for hostility and anger, but this may be because our brains flags such incidents as important to remember. However, there are as many, if not more, instances of genuine support and kindness within these digital spaces that we often overlook because they don't trigger our emotional brain as powerfully as negative experiences. Even a single negative comment can have a lasting impact, eclipsing a host of positive comments. Some individuals are more prone to dwelling on negativity and being in low power mode can exacerbate this. In such a state, content that hits our vulnerabilities will affect us more significantly, as we're emotionally drained and less capable of regulation. Regrettably, this is often the time we're most drawn to social media, using it as a coping mechanism.

If this discussion has resonated with you and you feel that you have pre-existing vulnerabilities that may be magnified by certain types of content on social media, it's essential to be more mindful of the type of content you expose yourself to. While body image content on social media has received significant research attention, you can apply the same considerations to any aspects where you might have pre-existing insecurities, be they academic performance, career aspirations or relationships. You should be mindful of the effect of your digital diet, just like you would your physical diet. Similar to how overconsuming unhealthy food can have negative effects on your physical health, recognize that the content you consume can significantly influence your thoughts, emotions and perceptions, particularly due to the interaction of your attention spotlight and social media algorithms. You can harness the power of meta-cognition: knowing how the attentional spotlight works and being aware of your negativity biases and vulnerabilities means you can challenge and adjust your attention spotlight when it starts to go awry.

The amplification effect of social media algorithms on pre-existing vulnerabilities is becoming more widely acknowledged. As a result, there are tools available that enable you to customize algorithms by explicitly instructing them not to display specific types of content, flagging them as harmful. If you encounter content that is particularly harmful or triggering to you, you should safeguard your mental well-being by removing it from your social media feed. While this may not eliminate all potential risks, it can substantially mitigate them, much like wearing a seatbelt reduces the risk of injury in a car crash. As a general principle, aim to engage with content that supports personal growth, presents diverse perspectives, and fosters a healthy mindset – use all available tools to tailor your social media feed accordingly.

Comparison

Contrary to popular belief, the impact of social media on our self-esteem is not universally negative. By pooling numerous scientific studies, we find that the overall association is either negligible or shows a small negative correlation.[144,145] A study examining how social media alters self-esteem found that 88 per cent of participants reported no effects, 4 per cent reported positive effects and 8 per cent reported negative effects.[146] Social media is often criticized for its downsides but rather than having a universal negative impact, it tends to have a targeted effect on individuals with specific vulnerabilities.

This pattern mirrors what we observed regarding mental health, which raises the question of why some individuals are more affected than others. Part of the answer may lie in our natural tendency to compare ourselves to the people around us. We are often advised to refrain from comparing ourselves to others on social media, recognizing that people tend to showcase a curated version of their lives. While this advice is well-intentioned, it can be overly simplistic. Engaging in comparisons is a common behaviour that provides us with a sense of where we stand in society and how we relate to others. Completely eliminating this behaviour is therefore not realistic, as we are all predisposed to comparing ourselves to our peers, both in terms of achievements and attributes.

Some individuals have a higher inclination to engage in comparisons and you might instinctively know if you're the type of person who tends to draw comparisons more frequently. This characteristic is known as having a 'high comparison orientation'. Rather than being inherently bad, the effect of these comparisons is complex. For example, comparing ourselves to others in similar circumstances can normalize our experiences and provide a sense of connection. These comparisons, known as lateral comparisons, are instrumental in fostering a sense of solidarity and support among individuals navigating similar life events or facing shared challenges. Many of my patients who have been diagnosed with the same condition have found solace and comfort, as well as valuable insights and practical support, by connecting with others in the same situation through social media platforms and comparing experiences.

When we engage in comparisons with others, we often evaluate our accomplishments, personal qualities, and social status.

The act of comparing ourselves to those we perceive as superior, known as 'upward comparisons', can elicit two distinctive reactions. For some, these comparisons can trigger negative emotions such as envy and feelings of inadequacy, leading to self-doubt. However, for others, upward comparisons can serve as a constructive tool. Observing others might help them objectively identify their own strengths, pinpoint areas for improvement, and inspire personal growth.[147] As always, the specific impact is contingent on individual predispositions and vulnerabilities. For instance, those experiencing depression may be more prone to the adverse effects of upward comparisons due to memory biases and negative self-perceptions, potentially intensifying feelings of low self-esteem and reinforcing negative self-beliefs. When it comes to body image, the frequency of unfavourable appearance comparisons someone engages in can be linked to dissatisfaction with their own physical appearance.[148]

On the opposite end of the spectrum are downwards comparisons – these are comparisons where we contrast ourselves with people we perceive as inferior. While these comparisons might initially seem beneficial, as they can temporarily boost our self-esteem by underscoring our own perceived superiority, they come with their own set of issues. For one, self-esteem that's largely based on external factors is inherently unstable, as it ebbs and flows depending on the people and content we encounter. Furthermore, this habit of seeking reassurance from downward comparisons can cultivate a fragile sense of self-worth and a sense of complacency, potentially hindering personal growth. Rather than striving to reach our full potential, we may find ourselves settling for a sense of superiority predicated on others' perceived shortcomings. Additionally, it's worth considering how this practice might affect the way we view or treat others, potentially leading to negative consequences in our social interactions.

Rather than dismissing comparisons altogether, it is more helpful to develop a nuanced understanding of how these judgements impact you, particularly your own inclination towards comparison. While social media may seem more candid than traditional media – and in some respects it is – the glimpses we get are still, to an extent, curated to showcase the best, and most enviable, aspects of people's lives. This presents an unfair comparison landscape, as we only see what others choose to share, a principle that extends to life in general. Additionally, as emphasized throughout this book, our phone usage habits can be the outcome of emotional triggers. This means we disproportionately scroll through social media during periods of emotional struggle or low power mode, intensifying the contrast between how we feel and the content that we see. At those times, our brain tends to overlook the moments of genuine joy and engagement we experience offline, away from social media, our attention spotlight magnifying this further by focusing in on the best and most desirable qualities we see while contrasting them with our own internal struggles and anxieties.

To navigate the complex landscape of social comparison, it's essential to foster balanced digital habits and draw upon a variety of emotion regulation techniques, as outlined previously. A helpful strategy is to shift detrimental upward or downward comparisons into lateral ones. When you find yourself making harmful comparisons, especially with people you don't know, make a concerted effort to find common ground. Negative upward comparisons, where another's success triggers feelings of inadequacy, can be reconfigured into opportunities for self-improvement. Extract lessons from these instances and use them as stepping stones towards your personal growth. Adopt an analytical approach – recall that the activation of the emotional brain prompts the executive to pay attention. Instead of centring your thoughts on the other person, delve into what this comparison is trying to tell you about yourself. Harness meta-cognition to decipher what this comparison reveals about your desires, values, or aspirations. What insecurities does it amplify? How does it align with your personal goals? It's important to understand that developing this form of introspection can be a more significant challenge for some individuals, and it often requires consistent effort and practice.

Be aware that doing this at times of low power mode, where your emotion regulation capacity is reduced and you are at higher risk of a negative internal narrative, will be more challenging. It is possible that some comparisons cannot be effectively approached that way. If you find yourself repetitively engaging in 'hate-following', or habitually viewing content or profiles that stir up negative emotions, consider this a red flag. Substitute this with people who inspire positivity, learning, or genuine enjoyment. Ultimately, the aim is that the digital environment should support your personal growth, not compromise it. Keep in mind that the content we engage with, and our interaction methods, can bring about tangible changes in our brains due to synaptic plasticity. Consider your brain as a VIP area, and exercise discretion in choosing the content and individuals you allow access to it.

Rather than fixating solely on social media, we should also recognize that these comparisons occur not only in the online realm but also in our offline interactions with the people around us. Many people will compare themselves with those in their immediate circles and, while online comparisons are transient, offline ones may be more persistent and potentially have more impact.[147] In any case, it is important to foster a balanced and healthy approach to self-esteem that is less affected by upward or downward comparisons with others. We need to focus on understanding our own brain with its intrinsic qualities and personal values in order to encourage individual growth and provide a more stable sense of self-worth, both in the digital and in the physical world.

Social media habits

If social media was universally bad, then it would follow that quitting would be universally good. But this is not what scientific evidence shows. A 2021 study randomly assigned 130 undergraduates into five different conditions: no change in social media use or abstinence for one, two, three or four weeks from all their social media platforms. The findings of this study showed no difference in well-being between those who continued to access social media and those who abstained for different lengths of time.[149] On the whole, when combining all the studies that have been done in this area, there are no clear overall benefits to going on a social media break.[150] Some studies even show that there are negative effects – namely a decline in life satisfaction, worse mood and increased loneliness – when social media usage is completely relinquished.[151]

This does not mean that all the research in this area produces the same results. A much-quoted study showed that deactivating Facebook for four weeks had a beneficial effect on users.[152] However, this finding could be attributed to the study being conducted during the 2018 American mid-term elections. The politically charged atmosphere of that time possibly increased the amount of digital stress among users, shifting the balance of the results so that people gained more benefit from disconnecting than they would have at a different time. It's a striking example of how difficult it is to disentangle the effect that social media has on individuals from the wider context that we live in and makes it impossible to create a blanket rule that applies to everyone.

The uniqueness of each person's brain, the complexity of their situation and the role social media plays in our individual lives mean that certain individuals under specific circumstances may see value in stepping away from social media. However, there is no scientific evidence that recommends that you quit social media completely unless you want to. That being said, should you want to remain on social media, an intentional and mindful balance should be struck to have the healthiest relationship possible. Many studies show that those who successfully regulate their social media use are able to avoid potential negative impacts.[153] Developing supportive habits is a key part of that, and will equip us with useful tools for regulating our use, while allowing us all the benefits of social connection. A lack of control over our habits means that constant phone checking starts to conflict with our goals, possibly displacing other beneficial activities and, ultimately, have a negative impact. Yet, it's important to understand that having an offline period away from social media does not necessarily help you build healthy digital habits. This is what makes it so easy to slip back into previous patterns of use after a break. For many, the optimal approach is to remain on social media but channel the energy that would have been used to go on a 'digital detox' towards building and then maintaining healthy habits.

Setting boundaries is a good way to start establishing more intentional use but we should avoid the knee-jerk response of simply limiting the amount of time we spend on social media. While this may seem like a straightforward solution, it can perpetuate the same disruptive habits by encouraging more frequent, albeit shorter, social media checks. As a rule, shorter and more frequent checks have the potential to be much more intrusive than longer but less frequent checks. Brief checks are really small actions and are thus more likely to become automatic habits rather than intentional actions. Despite their reduced durations, these habits can be as detrimental as spending extensive periods on social media, as they continually interrupt your focus and daily activities. Lack of control over these automatic checks can transform social media from a source of enjoyment into a source of frustration.

A more effective technique is to pre-commit by setting a finite number of checks of social media per day – this is something that was discussed earlier in Pre-commitment (Building Block 4), but I would like to expand on it here, as this technique is the most impactful on regulating social media use. The aim is to consolidate any brief social media checks into longer checks, thereby creating extended periods of social media free time in between. The exact number of checks established will vary from person to person, depending on their needs and preferences. You can apply the 80/20 rule (Building Block 7) which means that the majority of the benefits gained from social media are gained in the first fifth of checks with diminishing returns thereafter. I personally check Instagram up to twice per day, and manually log in – this is an effective hurdle to further structure my use. This works for me, but, rather than imposing this on everyone reading, I want you to consider what works for you, as you might prefer one long check each day, or want to start with three or four checks a day. I have in the past experimented with checking Instagram once per day for an entire month, but I found that I benefitted from the extra flexibility that two checks provided. Rather than an ambitious number, the key is to set a finite number of checks that you can stick to and still get what you need out of it.

The key thing about this approach is that it doesn't restrict the amount of time you spend on social media. Once I open Instagram, I am free to spend as much time as I want on it, therefore making this technique feel less restrictive. When repeated over time your brain will encode a new pattern, the number of external and internal reminders linked to social media will reduce and it will no longer be your go-to action for self-interruption. Using social media can become just like the many other finite things that we do daily, such as brushing our teeth, rather than a constant frustrating presence or a behaviour that we cannot control. Even if your job requires regular social media access, you can still develop a pattern of checking/posting that prevents it encroaching upon your other tasks, in a similar way that you might batch check your emails. For a pattern to be set and encoded into your autopilot brain, it needs to be recurring to fulfil the fourth piece of the Habit Puzzle – Repetition – and achievable enough to be executed even in low power mode.

The reason this technique is so effective is that, by having a limited number of checks, it means that you are less likely to engage in automatic and habitual behaviour. You are more likely to plan ahead and use it intentionally. After all, when you are limited to a few checks, you won't want to waste one of them on a quick glance that lasts a few seconds. Planning your checks ahead of time activates your forward-thinking executive brain which considers the long-term consequences and prevents you acting on autopilot. What's more, using social media in this way will actually make it more enjoyable. Rather than using the app for

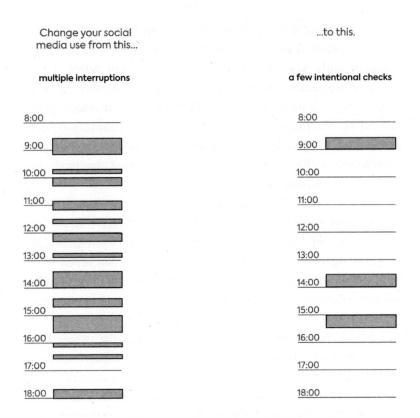

Change your social media use from this...

...to this.

multiple interruptions

a few intentional checks

How to Successfully Regulate Social Media Use: Instead of numerous mindless social media checks that become distracting, opt for a few pre-planned, intentional ones. This not only leverages anticipatory dopamine but also fosters a sustainable habit pattern that guards against overuse.

the app for social interaction or entertainment, the many brief unintentional checks we tend to make have become a form of avoidance – not a source of enjoyment. Remember, dopamine is anticipatory, and is released when we look forward to something. Acting on every impulse to check social media means we have no time for anticipation. By limiting yourself to a finite number of checks each day, and so refraining from constantly refreshing your feed, you will create anticipation and excitement for catching up with social media. This change in approach will lead to a more satisfying experience overall, which more than compensates for the reduced frequency of checks. This increased reward works in tandem with repetition to fulfill the third and fourth components of the habit puzzle, thereby helping to establish a new balanced digital habit. Interestingly, the opposite can also be true. Some people will find that, once their problematic habits are dealt with, they do not enjoy social media as much as they thought. They were just bound to this behaviour as a result of unsupportive habits in the autopilot brain.

The activities you engage in on social media significantly influence how you perceive your time spent there. While earlier research suggested active social media use, like posting, liking, and commenting, was more beneficial than passive browsing, further studies have found this isn't necessarily true.[154] Consequently, there's no prescribed 'correct' way to engage with social media – be it actively or passively. What's vital is that your social media habits align with your goals, and you're aware your interactions guide the algorithms to show more similar content. Habits of constantly checking social media as a result of distraction can fuel guilt and make the time feel like it's wasted. While taking an intentional break and choosing to use social media can be replenishing, being constantly distracted by it is very fatiguing for the executive. There is little value in just skimming through the latest news without any significant engagement. This does not necessarily mean that you have to actively like or comment on posts, but try to be fully present. Short bursts of opening an app, having a quick skim before quickly becoming uninterested, only to repeat the same process a few minutes later, might signal that you are in low power mode. Taking longer breaks to genuinely engage with the content, whether it's informative or entertaining, can enhance the quality of your social media experience, making your time spent there feel more valuable.

The abundance of information in the digital world, more than we can ever reasonably consume, means it is ever more important to be mindful of our purpose while online. If you are planning to spend time on social media, construct your feed to reflect this purpose as much as possible. For example, as an expectant parent, I felt more prepared by following the profiles of people who were discussing various facets of parenthood, ensuring a broad range of accounts but being selective about choosing content that would be personally helpful. If you're keen on mastering a skill, social media offers a diverse spectrum of knowledge and inspiration, from seasoned professionals to enthusiastic novices, to tap into. With so many professionals using social media, it is easy to find expertise on a range of issues as long as you choose your sources carefully. Not everything needs to be productive to be valuable. It's OK to be entertained or to look for inspiration and my feed includes some of that too. The main thing is to have a clear purpose for the time spent and work on building supporting habits that align with that.

Reward

We have formed habits relating to the way we check social media and the content we look at, but we have also formed behaviours surrounding our own digital contributions and interactions. Social media is exciting because of its novel and unexpected rewards – a comment or a message out of the blue, a flurry of likes, a new upload of photos. Broadly speaking, there are two types of rewards. Internal rewards are those which are able to fulfil psychological needs and to which we ascribe a powerful personal value. External rewards, on the other hand, provide a value that is dictated by others; its benefits are decided by people beyond us, by society. While using social media can be internally rewarding by creating a sense of creativity, self-expression or connection, we can also be heavily influenced by its external rewards, such as social validation or even monetary gain.

External rewards are not inherently bad but an overreliance on them is. Coupled with the unpredictability of social media, being over-reliant on digital validation metrics can fuel a particular type of problematic habit: constant refreshing. This often happens after we post something, and then refresh the app or feed multiple times in the hope that new notifications will appear. The reason this happens is, right before you post, there is an anticipatory rise of dopamine at the possibility of gaining those coveted external rewards. When the number of likes, or other metrics such as followers or views, is lower than expected, there is disappointment and dopamine drops (see illustration on page 103). You consider whether there has been a mistake. You refresh the app to double-check. And check again. This is like shooting a basketball – your brain wants to keep trying until you can succeed and end on a good note, by receiving the reward you expected.

Unfortunately, the external rewards of digital metrics are notoriously unreliable. Once a habit develops, someone may end up spending a considerable amount of time after posting merely refreshing and, rather than gaining a reward, this just fuels disappointment. You may have also established habits where you spend a disproportionate amount of time monitoring the myriad of online metrics that are provided rather than accessing other aspects of the app. This not only makes time spent on social media feel wasted, as a single check would suffice, but also increases the number of problematic short checks which become intrusive. The headspace that is devoted to this aspect of the digital world can increase online vigilance and, therefore, become a source of digital stress.

If you find yourself struggling with these habits, it is essential to recognize that external rewards on social media are often based on luck and can be inconsistent. Instead, try to tune in on your internal motivators for posting. In one study, it was found that individuals with a strong sense of purpose in life were less influenced by social media likes in relation to their self-esteem.[155] This indicates that having internal drivers and goals that are personally meaningful can act a protective factor, diminishing the impact of social media validation on your overall sense of self-worth. You could also consider developing your own reward system, providing yourself with a reward that you can control and so reducing your reliance on the external rewards of others. This should be based on effort rather than results, which can be unpredictable and demotivating when they do not meet your expectations. By doing this, you also cultivate more intentional habits around posting on social media as you need to think more carefully about what material you are sharing, and why.

While there is much more that could be said about the effects of social media on the human condition, I have tried to condense this into topics which you, the reader, can directly apply. It is clear that social media has both positive and negative effects, and we should avoid a simplistic approach that either exclusively damns it or celebrates it. While I provide practical suggestions to mitigate negative effects, equipping you to adeptly navigate the digital sea, it does not absolve society or social media firms from their responsibilities. We need to continue scrutinizing our technology with a view to increasing digital safety, much like how seatbelts, airbags and speed limits have improved road safety. I hope to see comparable measures taken for social media, such as the provision of a universal set of tools designed to help individuals establish healthy digital habits and promote digital well-being.

Social media practical

Rules for Healthy Social Media Use

Remember:

- The content you consume may have more effect than the time you spend on an app.

- What you are naturally inclined towards seeing may not be the same as what is beneficial for you, given how your attention spotlight functions.

- Skim-reading or mindless browsing means that you will be less likely to remember the content, leaving you with the sense that your time is not well spent.

- Getting distracted, even while on social media, might be a sign that you are on low power mode and that you need a break.

Action:

- Ensure you are following people who are educating, entertaining you or elevating your experience in some way.

- Do a digital declutter where you remove/unfollow/mute accounts which have a negative effect.

- Tune into how you are feeling when accessing certain types of content. To be more objective, consider whether you would like a family member/friend to feel that way.

- Customize algorithms by only liking/commenting on useful content and familiarize yourself with how to report/flag damaging content on the social media platform(s) that you regularly use.

- Prioritize intentional interest-driven browsing over mindless browsing.

Regulate social media checks

- Set a finite number of times for checking social media, condensing a large number of short checks into defined, longer ones.

- To decide on the exact number, consider how often you currently check social media per day. Try applying the 80/20 rule (see page 135), which suggests that by reducing your checks to one fifth of your current rate, you could still retain the majority of the benefits.

- You do not need to be overly ambitious. Choose a number you can consistently meet, because the reward of hitting your goal powerfully saves the habit. You can always reduce this further later if you would like to.

USE YOUR BUILDING BLOCKS

- Every time you are tempted to do a quick check, apply the Five-minute Rule (Building Block 1) to activate your executive, provide perspective and break apart the first two pieces of the habit puzzle.

- Relocate your social media apps on your phone to the last screen or a folder to Insert Hurdles (Building Block 3). Boost this effort by manually logging out each time and manually inputting your password; for an even bigger hurdle, utilize the pause provided by two-factor authentication.

- As a form of Pre-commitment, delete the app when you reach your designated number of checks and reinstall it the next day (Building Block 4).

- Substitute social media checks (Building Block 9) with activities like a short walk, stretching, gardening, word games, messaging a friend, or even organizing the photos on your phone.

Tackling External Validation

Strategy

- Hone and question your internal motivators to become less reliant on external validation. What are your specific intentions behind posting?

- Think about your purpose when you're on social media – are you improving any skills or learning new information?

- Practise rewarding yourself for consistent and intentional posts rather than unpredictable and unexpected results.

If you find yourself excessively monitoring the metrics on your posts:

- Apply the Five-minute Rule (Building Block 1) when you get the urge to check your likes or comments.

- Log out of social media after posting to Insert Hurdles and break the constant refreshing habit (Building Block 2).

- Planning a treat after posting can serve as an alternative and more predictable reward. It has the bonus of reinforcing your new habit through Temptation Bundling (Building Block 10) and providing a necessary distraction.

- Use Pre-commitment (Building Block 4) and schedule a post for a time that you will be busy so you cannot check the response until later.

Navigating social comparison

- Shift comparisons: Consciously transition from detrimental upward or downward comparisons to lateral ones. Find common ground and use these comparisons for personal growth rather than sources of self-doubt.

- Be analytical: When your emotional brain prompts the executive to pay attention, delve into what the comparison is trying to tell you about yourself.

- Guard your mental space: Not all content deserves your attention or analysis. If you encounter content that is particularly harmful or triggering, remove it from your social media feed.

- Employ mindfulness: It is possible that social media has become a source of digital emotion regulation when you are in low power mode. Use a variety of emotion regulation techniques from the previous chapter to maintain balance.

★ Remember: every step towards healthier digital habits counts, and progress may vary from person to person. Take this journey at your own pace and don't forget to celebrate small victories along the way.

How to deal with negativity online

- Your emotion regulation resources are finite. Establish boundaries and do not engage in anything that is too emotionally draining if you do not have capacity.

- Negative comments online are particularly emotionally draining. It is a natural instinct to want to reply to negative comments, and defend yourself, because, after all, the thought of it going unchallenged feels equivalent to confirming it. Give your brain space before replying to negative comments by taking a few minutes to think.

- Don't let the negative comment change your behaviour. If you would usually answer then do, if you wouldn't then don't.

- It is important to guard your emotional energy in the online world so truly consider whether engaging with a comment is worth the energy investment. If you respond, there will likely be further comments so you will be signing up to continued energy commitment.

★ Important: If something bothers you in the virtual world, ensure you speak about it in the real world. Our brain has a way of magnifying things when kept inside and it really helps to have someone else's thoughts for perspective. For example, once said out loud, a lot of trolling comments sound ridiculous and lose their power.

13

The Future

As we near the end of this book, I hope you've come to see that your smartphone isn't as sinister as it has perhaps been portrayed, and that your understanding of what is happening in your brain when you use it can help you avoid many pitfalls. Yet being the future-oriented creatures that we are, many of us worry about how smartphones might affect us in the long term. Will they waste our time, harm our brains, and cause problems for the next generations? I would like to conclude this book by sharing some neuroscience-based insights and knowledge that I hope will help alleviate fears about the future, however uncertain it may be.

The future you

Beyond building a time machine, learning from those who have come before us often provides the best insights into our future. I consider myself incredibly fortunate to frequently have these experiences, both personally and professionally. From treasured moments spent absorbing my grandmother's century-spanning tales of a transforming world, to life lessons instilled by my parents, to narratives shared by patients I've been privileged to care for, each offers a window into different future possibilities. When I hear their life experiences recounted, I'm reminded of a quote by Danish philosopher Søren Kierkegaard: 'Life has to be lived forwards but can only be understood backwards.'

The key difference between you, engrossed in this book right now, and the future version of yourself, lies in the memories you forge and retain along your life's journey. It is these memories that will form the narrative of our lives. I've always been intrigued by how our brains process and store memories, and this personal interest culminated in some of my scientific research. A crucial detail many people don't realize is that our brains store different types of memories, each playing a unique role in shaping our experiences. There are semantic memories – or factual memories – and episodic memories, which is our recollection of specific events, many of them highly personal and unique to us. For instance, knowing that Paris is the capital of France is a semantic memory, while recalling a delightful weekend spent in Paris is an episodic memory.

Working hard, constantly studying, answering messages, finishing your to-do list, getting your inbox to zero are all activities that build semantic memories. The key difference between these two types of recollection is that semantic memories lack the 'when' aspect that episodic memories have. While you may know many facts, you may not remember when you acquired them. And a lack of episodic memories can make it seem like your life is a bit of a blur. Episodic memories hold the power to transport us back to cherished moments, reminding us of the joy, laughter and unique moments that make our lives rich and meaningful.

Many of us have experienced how busy times, like an intense period at work, studying for exams, or even the pandemic lockdowns, can blur our sense of time and memory. It's a phenomenon I've personally experienced. Besides the distinct memory of picking up *The Man Who Mistook His Wife for a Hat* – a moment that steered my career path and formed the backbone of this book's opening chapter – the rest of that academic term seems fuzzy. It was a period filled with countless hours of rigorous study, which left me with an abundance of semantic memories, but very few moments that stand out in my personal memory reel.

While I don't regret the long days in the library, or the time spent crafting this book for that matter, living a fulfilling life demands a harmonious blend of embracing joyful moments and committing to hard work. It's about creating a rich tapestry of both semantic and episodic memories. While self-control and productivity are often highly coveted attributes, we must remember to allow ourselves the opportunity to form enduring episodic memories. These memories become the threads that weave our life stories, evoke immense joy when we relive them and become cherished anecdotes to share with others in the future. Studies have shown that experiences often contribute more to our happiness than material possessions, provided our basic needs are met and we have enough to get by.[156] In line with that, creating chances to foster episodic memories is a key part of my own personal philosophy and I think it is one of the most overlooked ways to invest in and care for your future self.

How then, do phones and technology affect our memory formation? Habitual phone use tends to produce fact-oriented semantic memories which lack the richness of personal experiences. In reality, mindless scrolling and aimless device interactions might even fail to create any memories at all. This is especially the case when phone usage becomes excessive. The act of scrolling and binge-watching may lead to a disproportionate amount of content consumption without pauses for our brain to think, analyse, or reflect on what we have seen. Consequently, these experiences often don't succeed in creating memories which are robust enough to be stored for future use. It's not uncommon for people to remember less of what they've seen due to an information overload. As a neuroscientist, this outcome isn't surprising. We know that spaced repetition – revisiting information slowly and periodically – significantly outperforms a single intense study session in cementing memories,[157] and the same principle is true whether we're preparing for an exam or browsing our social media feed. We can counteract this through mindful use of our devices – intentionally selecting the content we consume, anticipating it, and savouring it. For instance, whether we are reading an article, watching a film, or browsing social media, taking the time to reflect and absorb the content can lead to a deeper understanding and a more memorable experience – this can improve the quality of our semantic memories. By cultivating positive digital habits, we can also exploit the vast potential of technology to bring joy and excitement into our lives, just as my grandmother often recounted in her stories. Consciously engaging with our devices and using them as tools to connect with loved ones means that we can create meaningful moments. In this way, we harness the power of technology to create a balance of episodic and semantic memories and enrich our lives.

The inner workings of memory storage are truly fascinating. Imagine a new memory as a piece of mail that lands in our memory inbox, the hippocampus. Over time, particularly through the process of sleep, this 'memory mail' gets moved into long-term storage in various parts of the brain. Now, often when we struggle to recall something, it's not due to a failure in this storage process. Evidence for this is when a tiny clue, like the first letter of someone's name, suddenly brings the memory rushing back. This indicates that the memory was indeed stored properly, but we had trouble accessing it and bringing it to the forefront of our mind – something we call a 'retrieval error', a common hiccup in the recall process which happens to all of us due to the vast amount and complexity of the information our brains store.

Whereas a forgotten factual memory can be easily looked up, retrieving episodic memories – those personal, unique moments – isn't as straightforward. Sometimes, we need the help of others to fill in the gaps or to piece together details, like the charming restaurant that we visited while on vacation or the breathtaking view from a mountain we hiked. This is where our smartphones can be invaluable as memory aids. Today, we capture more of our lives than ever before. Just a quick look through my phone's photo gallery sparks a deluge of forgotten memories. My phone also has the ability to showcase randomly selected photos of my children on the lock screen – this not only brings a smile to my face, but it also reinforces the neural pathways connected to those memories, letting me revisit those moments more often than I would otherwise. If I'd had a smartphone during my first year of medical school, my camera roll would be full of joyous memories from that time, all just a scroll away, making those moments easier to relive.

Indeed, our phones are invaluable tools for preserving memories, but it's crucial not to let them disrupt us during significant occasions. It's perfectly fine to capture these moments, but over-documenting might compromise the richness of the experience itself, diminishing the vibrant details that will enter our memory inbox. A useful practice is to use our phones as aids, not as distractions. This could involve taking a single photo, rather than multiple, and saving the task of posting it on social media for later. This practice not only helps maintain your presence in the moment, but also gives you a chance to relive it later. Moreover, cultivating good sleep habits – technology-related or otherwise – promotes the consolidation of our memories. Sleep facilitates the transfer of these experiences into our brain's permanent storage, enhancing our capacity to recall – and relive – these cherished moments.

Another benefit of our smartphones is that they provide a unique form of escapism, allowing us to momentarily immerse ourselves in different worlds and explore perspectives beyond our own. This enriches our life but it's crucial to strike a balance. In navigating these digital landscapes, we must never lose sight of living and engaging with our own lives. After all, how we spend our time and attention today directly shapes our future selves. Mindfully cultivating digital habits that are in harmony with our personal goals and aspirations is an investment which means we can utilize technology as an enriching tool that amplifies our purpose, rather than a distraction that deviates us from our path. This becomes a key strategy in caring for our future selves, helping us evolve into the individuals we aspire to be.

Future-proof your brain

'For this invention will produce forgetfulness in the minds of those who learn to use it, because they will not practise their memory.' This quote so aptly underscores our ongoing concerns about the impact of technology on our brain, that you may be surprised to learn that it is attributed to the philosopher Socrates, in around 400 BC, and that he was actually referring to writing.[158] To delve deeper into this concept, we need to first explore the neuroscience behind memory and forgetfulness.

Having spent a lifetime building our memories, to be able to enjoy them it is important to protect our brain's machinery. The case of H. M., in Chapter 8, demonstrates the consequences of damage to the memory inbox: an inability to create any new memories. However, for most people, forgetfulness doesn't result from surgery or trauma. The most common cause is the mis-folding of proteins inside our cells. During sleep, some of these misfolded proteins are cleared but, as we age, they can gradually accumulate. Once they reach a critical threshold, they begin to disrupt the internal machinery of our neurons, leading initially to dysfunction and eventually to permanent damage. This process is known as neurodegeneration. The specific protein involved and the affected area in the brain determine the particular disease that a neurologist will diagnose. For instance, damage to the substan-tia nigra, the crucial dopamine-producing region that modulates our movement, results in Parkinson's disease. When it comes to forgetfulness, the most commonly diagnosed form of dementia is Alzheimer's disease. In this condition, two proteins, beta-amy-loid and tau, accumulate outside and inside the neurons in our memory inbox, forming what scientists refer to as 'plaques' and 'tangles'. This accumulation disrupts the neurons' inner workings and hampers their ability to communicate effectively.

The effect of Alzheimer's disease on the memory inbox is something that I often encounter in clinic. Patients in the early stages of the disease will often report difficulties remembering recent events, while their long-term memories, already in permanent storage, remain intact. Regrettably, as the disease advances and the misfolded proteins extend to other areas of the brain beyond the hippocampus, they disrupt both long-term memories and other cognitive functions in the process. Extensive research efforts are ongoing, but reversing neuronal damage is proving to be a very difficult feat. However, in 1988, a remarkable discovery was made when a group of scientists conducted a study on 137 post-mortem brains while aiming to assess the levels of misfolded proteins. To their surprise, the researchers found that a proportion of the brains had a similar level of plaques to that you would expect from someone suffering with Alzheimer's disease. Yet these people had never reported any memory issues while they were alive. In fact, their mental performance was just as good, if not better, than those who did not have any brain pathology.[159]

Why did those particular individuals fare better than expected despite the presence of misfolded proteins? It turns out that their brains had a larger size and a greater number of connections. Despite some neuronal damage, these people had enough reserve capacity for their brains to compensate and function normally, without displaying any symptoms. This is what is known as cognitive reserve, and has been supported by numerous studies.[160] Think of cognitive reserve as a savings account that you gradually build up over time. Similar to how having savings protects you from unexpected financial burdens, cognitive reserve can safeguard against memory decline.

Some decline in how our brain functions is inevitable with age. Studies show that there is a 0.2 per cent brain volume loss yearly after the age of thirty-five and this slowly accelerates to greater than 0.5 per cent in those over sixty.[161] Building cognitive reserve is the most effective way to future-proof your brain. Education and engaging in intellectually demanding occupations that involve complex thinking are closely associated with higher cognitive reserve. The brains of those who take part in these activities show an increased number of connections and greater overall thickness. Throughout this book, we have explored the concept of 'cells that fire together, wire together', highlighting how our brain forges strong connections based on the actions we engage in. Conversely, 'if you don't use it, you lose it' holds true for the brain as well. Neglecting certain activities can lead to underdevelopment in the corresponding brain regions, ultimately affecting our cognitive reserve in the future.

Socrates' concerns about writing were therefore not entirely baseless. The advent of writing brought about a decline in overall memory capabilities compared to his time when individuals could recite lengthy passages from memory alone. However, this cannot be taken in isolation and it would be unfair to dismiss the profound positive impact that writing has had on our brains. Writing exposes us to a wealth of ideas beyond our immediate surroundings, broadening our intellectual horizons and stimulating our thinking abilities, such that the benefits to our cognition vastly outweigh this decline. We do not need to revive near-extinct practices – such as recalling huge reams of text – to keep our brain active. While our memory capabilities may not be exercised in the same way, reading and writing still keep our brains actively engaged. Indeed, the connections in your brain will have undergone significant changes as a result of reading this book. Writing is a powerful tool for building cognitive reserve and future-proofing our brains. Rather than fuelling our forgetfulness, it is safeguarding against it.

Socrates' apprehensions, however, do echo some of our current fears about how we use technology, so we must now apply the lessons of the past to the future. One common worry surrounding our phone use is its potential to diminish our cognitive abilities. For instance, growing up in the pre-smartphone era, I had the ability to remember multiple personal phone numbers. However, aside from my childhood home whose phone number remains firmly embedded in my memory, I can now only recall two numbers – my own and my husband's. And I had to deliberately memorize the latter during the late stages of my first pregnancy to ensure that I would be prepared for an emergency. This is something that many people reading this will relate to. Yet the fact that

I no longer retain a multitude of landline numbers in my mind does not indicate a failing memory. It simply reflects a shift in how my brain is being utilized. I now allocate more of my cognitive resources to recognizing subtle differences in brain scans, remembering names of obscure medical conditions and recalling various drug dosages. This is because our memory system prioritizes and preserves information that is regularly used, while allowing less frequently accessed data to fade away. For instance, as a doctor, I find that pager numbers, codes and passwords are ingrained in my memory when in regular use – forming memories equivalent to the frequently-used landline numbers – but these memories are quickly replaced by new ones when I switch hospitals, as the previous set of information becomes less relevant.

If you, like me, grew up in the pre-smartphone era, you may notice a distinct shift in how your brain has adapted to the change in information management. The necessity to remember a multitude of phone numbers has been made redundant by our smartphones, and we tend to remember only the ones we frequently use, such as our own, due to the necessity of repeated recall. Consequently, our cognitive resources may have been reassigned to tasks necessitating a similar frequency of recall, such as remembering complex passwords, email addresses, two-factor authentication codes, or even social media handles. We also need to bear in mind that our view of the past is often tinted by nostalgia. We are prone to overemphasizing the amount of information we could retain before the proliferation of smartphones, conveniently overlooking our reliance on physical aids like phone books or other paper records. The truth is our brains have used external tools to manage less frequently used information for a long time – since the advent of writing in fact. It's just that the nature of these tools has evolved over time.

Just as exercising builds cardiovascular fitness, challenging our brain mentally builds cognitive reserve. A fascinating example of this can be seen in the brains of London taxi drivers who must pass a rigorous test – 'the Knowledge' – about the city's streets. Brain scans have revealed that these drivers have larger hippocampi.[162] In addition to serving as the memory inbox, the hippocampus also has a powerful role in spatial navigation and this is why becoming disorientated and getting lost is one of the earliest symptoms in people with Alzheimer's disease. In taxi drivers, the demanding nature of constantly navigating a complex cityscape directly impacted the size of this crucial brain region.

This is where some of the concerns regarding technology may have some merit. For example, it is possible that an overreliance on technology for navigation, such as the GPS on our phone, will not have the same effect on hippocampal growth, possibly resulting in diminished cognitive reserve and a higher incidence of dementia as we age. However, it is critical to underline that this is currently just a theory – a Socratic worry, if you will. There may well be compensatory benefits, much like the cognitive benefits we have seen associated with the advent of writing. To conclusively determine the impact of technology on cognitive reserve, we would need long-term studies spanning several decades, an advantage of time that we simply do not have at present.

It's worth noting that despite the theoretical concerns, our digital future isn't necessarily a bleak one. Quite the contrary; there are multiple promising indications suggesting potential benefits. For example, remaining socially active is a potent source of brain stimulation that builds cognitive reserve,[163] something that is made easier through technology. A variety of studies show that social media offers several positive consequences for older adults, through opportunities for social engagement, reducing feelings of loneliness and thereby enhancing mental health.[164] Interestingly, a study even showed that social media use can help protect against the decline in executive function that naturally occurs with aging.[165]

As echoed throughout this book, the majority of the negative effects associated with phones stem from problematic and excessive use of technology, so any negative outcomes, including the theoretical impact on cognitive reserve, aren't inevitable and can be mitigated by employing the supportive digital habits recommended in this book. It is not about avoiding technology altogether but rather using it mindfully and purposefully. The main concern lies in overreliance, not simply utilization. Mindlessly scrolling through low-effort content can be beneficial for mental breaks, especially when our brains are fatigued as a result of engaging in cognitively taxing work. However, when we are not fatigued, it is important to avoid developing habits that prevent us from challenging ourselves.

The fundamental principle to remember is that whenever you challenge yourself mentally and strive for improvement, your brain will undergo positive changes. The source of the challenge, whether technological or non-technological, is of lesser importance. Exposing yourself to new ideas, learning a language, or playing problem-solving games all contribute to the development of cognitive reserve – and many of these are often made more accessible through our phones. We should all seek out mentally challenging activities and incorporate a diverse range of them to ensure a well-rounded cognitive workout.

To start integrating cognitive reserve habits, you can utilize the building blocks from your toolkit: deliberately set aside a dedicated time each day to establish a domino habit (Building Block 8); or substitute an already-existing activity with one that will build cognitive reserve (Building Block 9). Alternatively, you can be opportunistic and seize moments throughout your day to engage in cognitive challenges. Rather than immediately relying on your phone or the ease it provides, you can employ the Five-minute Rule (Building Block 1) to harness the power of your mind first. For example, take a few minutes to mentally plan a route before looking up directions, perform some mental arithmetic before using a calculator, or try to recall information from memory before resorting to an online search. The mental effort you put into these tasks, regardless of whether you arrive at the right answer, actively engages and stimulates your neural pathways, driving positive changes.

It is crucial to implement this advice with kindness. There will be times when your executive brain is fatigued and operating in low power mode and, during those times, your brain will require a break rather than additional mental challenges. Recognizing when your brain needs a rest is also essential for maintaining its well-being. Pushing yourself relentlessly without allowing for adequate recovery can lead to mental exhaustion and diminish cognitive performance. It is perfectly OK to take breaks and engage in so-called mindless activities, be it scrolling or something non-technological, to provide that much-needed recovery. By becoming aware of and reflecting on your habits, you can ensure that scrolling doesn't become your only default setting, because as we've learned, habits have the power to pre-define our options.

Digital natives are the future

Taking responsibility for yourself and your own decisions is one thing, but assuming responsibility for another life is an entirely different matter. This is something that I face in my day-to-day job as a doctor. But even that is vastly overshadowed by the weight of responsibility that becomes apparent when you cradle a tiny, vulnerable human in your arms and realize that their well-being and survival are entirely dependent on you. Their every need, from nourishment to safety, relies on your diligent care and attention and parents are constantly reminded of the profound impact their actions today may have on their offspring's development and future.

This level of responsibility is daunting and parenting is scary – and arguably parenting in the digital era is even more so. The parents of digital natives today – particularly given they did not grow up with ubiquitous phone possession – are navigating uncharted territory and all the challenges that poses. Unfamiliarity powerfully activates our emotional centre: a study scanning participants' brains showed increased activity in the amygdala when they were presented with unfamiliar images, which subsided once these images had become recognizable.[166] We all mount an instinctive fear response in the face of the unknown. I experience this myself as a parent to two young daughters. Fear serves as a protective mechanism: the emotional brain is providing an alert to the executive brain urging us to be cautious and attentive in the welfare of our children, a feeling that resonates with most new parents. It is how we react to this fear which is important. We need to acknowledge and respect it, rather than dismiss or run away from it.

A commonly used tactic of scaremongering headlines is to exploit the fear you feel to capture your attention. When the emotional brain becomes active, our brain navigates the attention spotlight to what caused this to happen. Parents, when faced with doom-laden headlines, should remember that negative claims tend to oversimplify the complex dynamics of families and ignore broader socio-economic factors that might contribute to the challenges faced by future generations. The hype tends to boil down all problems to a single factor: smartphone use. The phone has become a symbol of the potential harm your child might experience online. This overly simplistic interpretation brings to mind early studies that found a correlation between having regular family dinners and a handful of positive outcomes, such as improved academic performance and reduced substance abuse among children. This narrative became so strong that the family dinner was even touted as a method to combat drug abuse.[167] However, while family meals can have benefits, it's not the act of sharing a meal that reduces substance abuse. A closer look revealed that the act of eating together served as a proxy for other positive family dynamics like good communication and support. In fact, when studies began to account for these additional variables, the correlation weakened considerably.[168,169] Thus, it's crucial not to oversimplify these complex issues or attribute too much blame to any single factor, whether it's the absence of family meals or smartphone use. It's more about the quality of our interactions during these activities as well as what's happening in our lives outside of them. A loving family isn't just about eating dinner together, much like understanding the impacts on children goes beyond measuring screen time or the age they first use a smartphone.

What are so often missed in the narrative are the context-specific factors. Studies show that children with more highly educated parents and those with higher incomes are less likely to own phones at a young age and, in general, children attending more socially disadvantaged schools are more likely to have phones.[170] The reason this is the case may be because they are also navigating other challenges – living in unsafe neighbourhoods, having limited indoor living space with little to no outside space for playing, and shouldering responsibilities prematurely. The increase in technology usage could also be a symptom of overwhelmed parents who themselves are struggling either mentally, physically, or financially. While families with more resources may have the ability to research concerns regarding smartphone use, and to consider guidelines about how they are used and introduced at home, those who are less privileged may be focused on basic survival, which inevitably affects all aspects of their well-being, including their mental health. Current research on the effects of smartphones on children is far from conclusive. When you come across investigations into this issue, or more of those sensational headlines, approach them with a critical mindset. By keeping the broader societal context in mind, we can steer clear of over-simplifications that fuel stigmatizing narratives or unfairly blame individual family decisions.

Having this broader perspective can temper our own instinctive fear response. As we learned in the chapter on Mental Health, we have the power to guide our children's emotion regulation by distinguishing what is genuinely frightening from what isn't. Hence, we should not let our actions be solely driven by fear. Amplifying fears, whether related to food, academic performance or smartphones, isn't productive. Instead, we should use the emotional alert triggered by fear as a catalyst for positive change. Your fear likely reflects a deep investment in your children's well-being, which of course is a good thing, and it can be a great motivator to learn more about how to safeguard and improve that well-being. Perhaps that's why you're reading this book: to turn fear into knowledge, growth, and positive action.

While parenting in the digital age poses new challenges, the task of teaching our children about potential dangers and guiding their behaviour is something we are well equipped to do. In the same way we teach them to handle scissors safely, cook responsibly, and cross the road cautiously, we must also equip them with the skills to navigate the digital world. Just as the physical world can pose risks, the digital landscape has its own set of dangers but rather than instilling anxiety, or advocating avoidance, our focus should be on educating them about safe digital practices, just as we would teach them about road safety.

The key difference between technology and the other dangers described here, however, is its habit-forming nature. In 1890, the psychologist William James wisely wrote, 'could the young but realize how soon they will become mere walking bundles of habits, they would give more heed to their conduct while in the plastic state' – a remarkably forward-thinking concept at that point in time, given how little was known about the brain. As I have often stated, banning technology or setting overly strict boundaries might seem like a good idea, but cutting some-thing out does not always provide an effective safety net. We can't shield our children from technology forever or pretend it doesn't exist. This is neither a practical solution nor does it form those good digital habits. Moreover, being too dictatorial in our approach to phones can lead to frustration and strain the parent-child relationship. Children will likely find ways to engage with technology if they feel overly restricted, which comes with the danger that they may not want to tell you if there is a problem. As parents, our ultimate goal should be to raise future adults who can navigate the digital world independently. Digitalization will continue and be an integral part of our future so, instead of avoiding it, or solely relying on strict external rules, we should help our children build their own internal ones. We have the opportunity to shape our children's developing brains, helping them encode good digital habits into their autopilot in a way that will benefit them throughout their life.

This, however, doesn't mean that we should be overly permissive. The guidance we offer hinges on our children's age. Younger kids, with their still developing executive brains, often grapple with long-term thinking and self-control. Handing them unfettered access to technology equates to giving them an unlimited stash of sweets. We should instead apply the principles of the four-piece Habit Puzzle, which are universally applicable across different aspects of life including technology and food habits. We can guide our children in developing good habits around activities like watching TV or using a tablet, while allowing them a sense of control where appropriate. For instance, when my eldest daughter was just two, I started giving her the responsibility of turning off the TV. I would inform her that our viewing time was nearing its end but allowed the autonomy to determine the precise moment to push the off button. As she got a bit older, around age 4, I started helping her set up her own implementation intentions (Building Block 4). We'd chat ahead of time about how long she planned to watch and how she'd know when it was time to wrap up her screen time. These small acts aimed to help her develop her self-regulation skills. Instead of being fully reliant on my executive power, the goal was to help her develop her own. If she found it challenging, I would step in to help, offering support instead of criticism, and over time, she improved. Adopting this method, similar to establishing habits in other aspects of their lives like bedtime routines or practising good hygiene, paves the way for the development of healthy phone habits as they grow older. Don't be disheartened if things don't go as planned immediately. It's essential to remember that repetition is a key piece of the Habit Puzzle, and much like it takes years for children to get into the groove of brushing their teeth independently, the same applies to technology use. It's entirely normal for our children's evolving brains to be somewhat inconsistent and to push the boundaries, but we need to maintain a consistent, confident and calm approach.

As our children mature and their executive brain develops, it becomes increasingly important to respect their viewpoints and allow their perspectives to inform our decisions. Rather than imposing rules without context, we should collaborate with them to help them build supportive digital habits and, with the aid of this book, help them understand how their brain functions. This means knowing how their brain functions, including how habits are encoded in their autopilot, an awareness of when their executive enters low power mode, how the emotional brain works and the mechanics of sleep. The ultimate goal is to create an atmosphere where technology use is a topic of conversation, not a cause for contention. Remember, we're not just setting boundaries – we're setting the stage for habits that will last a lifetime.

This attitude to phone use aligns with a broader ethos that supports parents fostering resilience, cultivating self-esteem, and helping our children discover their purpose in life, rather than simply shielding them from challenges. Avoidance of all digital content is a short-term solution but teaching them to recognize which digital content is helpful – rather than harmful – is a long-term skill that they will inevitably need at some stage. Developing a stable sense of self-esteem is crucial in ensuring that external factors, whether that is social media likes or other factors such as academic performance, do not disrupt their well-being. Encouraging our children to discover their purpose in life is vital: a strong sense of purpose has been linked to reduced impulsivity and a decrease in the pursuit of short-term rewards.[171] While doing our best, it's important to remember that mental health issues are influenced by a complex interplay of various factors and should not be attributed to smartphone use alone. As parents, we should avoid blaming ourselves or our children for any challenges that may arise and instead seek support.

A big part of parenting our digital natives is modelling good digital habits ourselves, which can be challenging. Parenting is an intense and demanding role, and some parents find themselves in unsupported situations. Some may reach for their phones as a form of escapism, and they may use technology to occupy their children, potentially facing a great deal of judgement for doing so. Instead of resorting to a reductionist and simplistic approach that blames parents for their struggles, it is essential to consider the underlying reasons why some parents find it difficult to cope. Many parents struggle with access to affordable childcare, limited parental leave, or balancing demanding work schedules with raising a family. The pressure to juggle all these responsibilities with little support leaves many parents stretched thin, turning to technology to fill in the gaps or to multitask. What's crucial to remember is that it's all right to use technology sometimes. We can't run on empty all the time. To best look after our children, we need to make sure we're not constantly in low power mode ourselves. And if that means taking some time for ourselves with the help of technology, that's OK. The aim isn't to be perfect. It's to be mindful of our habits and to strive for a balance that benefits us and our children.

When considering our children's futures, we shouldn't forget that our brains are incredibly resilient and will adapt to incoming challenges. In my clinical practice, I'm privileged to frequently interact with younger generations, and I am not disheartened. They're dynamic, vibrant, involved in social causes, and far from the image often portrayed by media – that of passive zombies aimlessly scrolling on their phones. They're not just consumers of digital content but active participants, leveraging the digital world to voice their thoughts, share their creativity, and drive change.

The subject of children and technology is broad and complex enough to fill an entire book, so this brief exploration is not intended as an exhaustive guide. Instead, it aims to underscore one key message: by understanding how our brain works and taking control of our own digital habits first, we equip ourselves with the knowledge and experience to guide our children more effectively in the online world. As we navigate into the digital future together, remember that the cultivation of conscious, healthy habits is our shared responsibility and, indeed, our strongest tool. The principles we've explored throughout this book can be applied beyond our personal phone use, serving as a positive influence on our children's digital interactions.

The Future Practical

Future-proof your brain

Embrace these principles to enhance your brain's potential:

- Use your phone mindfully for memory preservation - smartphones can be a very useful tool for retrieving your episodic memory but balance this with being present in the moment. If you want to send images to friends or post on social media, do this later.

- Replace a problematic digital habit (Building Block 9) with reviewing photos or videos that you've taken to enhance memory retrieval and promote gratitude practice.

- Before using your phone's navigation, first attempt to plan the route mentally. As a further challenge for your hippocampus, sketch maps of frequently visited areas.

- Broaden your cognitive engagement by intentionally diversifying the functions you use on your phone. Substitute some habitual social media or news checks with cognitive reserve-promoting activities like language learning apps, educational podcasts, puzzles, or creative tools.

Guiding principles to help children

Developing supportive digital habits for children requires an individualistic approach, considering their unique needs, interests and maturity levels. I am not offering you a comprehensive or perfect solution but, based on the ethos of this book and what you have learned, here are some guiding principles to consider how you can approach teaching your children about using devices:

- Encourage supportive digital habits from a young age, such as allowing them to turn off the TV themselves or put a tablet away. The aim is to gradually enable – rather than restrict – them to make responsible decisions about their digital activities.

- Before engaging in digital activities, help your children plan ahead: 'How much are you planning on watching/ playing?' and 'How will you know when you are finished?' are great questions to help them engage their executive brain.

- Try to share the ideas from this book in a manner that children can understand, teaching them about their brain and how habits are formed.

- Support their executive brain by establishing guidelines regarding technology use collaboratively, involving them in the process whenever possible. Regularly revisit and adjust these boundaries together as their executive brain develops.

- Help them understand the value of a well-rounded lifestyle and that they should balance – and try to enjoy – both digital activities and offline pursuits such as exercising, socializing, reading or other hobbies.

- As with teaching a child anything new, recognize that forming habits takes time and repetition. Don't be discouraged if a strategy does not work immediately; keep reinforcing positive digital behaviours and provide consistent guidance.

- When online, encourage critical thinking about the content that they consume to keep themselves safe and treat others with kindness and empathy.

- Acknowledge any fear you feel and use that to make positive changes rather than knee-jerk responses. Instilling fear in your children or inadvertently encouraging covert use through prohibitive rules will not foster understanding or communication. You want to ensure that your child will be able to approach you with any issues, without the fear of being criticized.

Final Words

In this fast-paced digital era, where a wide array of content is competing for our attention, I am grateful to you for investing your time in reading this book – I wholeheartedly believe it is an investment that will pay you back many times over.

As a doctor, my primary responsibility is to help individuals with their well-being, while respecting their autonomy and guiding them in making decisions that are right for them. I therefore largely wrote this book focusing on you, as an individual, as if you were in my consultation room, sitting in the chair across from me, seeking guidance and support. Ultimately, my aim is to equip you with an understanding of your brain's inner workings and to provide you with the tools necessary to create supportive digital habits that align with your personal goals.

When working on a complex project such as a book, you hear stories of successful people who, in a bid to maintain focus and productivity, go to great lengths to disconnect – relinquishing social media passwords to assistants, using website blockers, or even going off grid to some remote location. I chose a different approach. By using my neuroscience knowledge of how the brain works and implementing the very same building blocks and strategies outlined in this book, I tried to develop a solid system of digital habits. This meant that, throughout the process of writing, I could remain connected and actively engaged on social media, regularly posting and interacting with readers. Instead of technology becoming a hindrance, it became an advantage, allowing me to leverage its benefits but without losing focus. This intentional connectivity enabled me to scrutinize the latest scientific data, as well as listen to the voices of potential readers, to learn from their feedback, and ultimately improve the content of this book.

While I have primarily focused on you, as an individual, it is clear that we don't exist in a vacuum. Our surroundings significantly influence us, and responsible tech usage goes beyond just personal actions. That's why I've made a concerted effort to extend the conversation beyond our individual brains, exploring how our living, working, studying, and parenting environments affect our habits. We need to acknowledge the pivotal role that governments, tech companies, and employers play in influencing this aspect of our lives. Listening to scientific experts, implementing supportive policies, and holding these entities accountable is a key part of cultivating an environment that promotes healthier tech usage.

That being said, finding effective solutions to address systemic issues will undoubtedly take time, and that is why this book exists – to provide you with actionable strategies and guidance that you can use to start reshaping your digital habits today. Similar to how a doctor would advise their patients on maintaining a healthy lifestyle even before any health-related laws are passed, this book is designed to help you to take control of your digital interactions, no matter what the external circumstances may be. With *The Phone Fix* at your side, you now have a guide to a future where technology serves your purpose and where you are equipped to rewrite your digital habits for life.

References

1. World Health Organization (2023).
2. Ofcom. *Online Nation 2022 Report* (2022).
3. Herculano-Houzel, S. The remarkable, yet not extraordinary, human brain as a scaled-up primate brain and its associated cost. *Proc Natl Acad Sci U S A* **109 Suppl 1**, 10661–10668 (2012).
4. Waters, J. Constant craving: how digital media turned us all into dopamine addicts. In *Guardian* (2021).
5. National Institute of Drug Abuse. Drugs, Brains, and Behavior: The Science of Addiction (2022).
6. Bowman, N. D. The rise (and refinement) of moral panic. In *The video game debate: Unravelling the physical, social, and psychological effects of digital games.* 22–38 (Routledge/Taylor & Francis Group, New York, NY, US, 2016).
7. Aarseth, E., *et al.* Scholars' open debate paper on the World Health Organization ICD-11 Gaming Disorder proposal. *J Behav Addict* **6**, 267–270 (2017).
8. van Rooij, A. J., *et al.* A weak scientific basis for gaming disorder: Let us err on the side of caution. *J Behav Addict* **7**, 1–9 (2018).
9. Petry, N. M., *et al.* An international consensus for assessing internet gaming disorder using the new DSM-5 approach. *Addiction* **109**, 1399–1406 (2014).
10. Przybylski, A. K., Weinstein, N. & Murayama, K. Internet Gaming Disorder: Investigating the Clinical Relevance of a New Phenomenon. *Am J Psychiatry* **174**, 230–236 (2017).
11. Bowman, N. D., Rieger, D. & Tammy Lin, J. H. Social video gaming and well-being. *Curr Opin Psychol* **45**, 101316 (2022).
12. Kardefelt-Winther, D., *et al.* How can we conceptualize behavioural addiction without pathologizing common behaviours? *Addiction* **112**, 1709–1715 (2017).
13. Zippia. 20 Vital Smartphone Usage Statistics [2023]: Facts, Data, and Trends On Mobile Use In The U.S., Vol. 2023 (Zippia.com, 2023).
14. Panova, T. & Carbonell, X. Is smartphone addiction really an addiction? *J Behav Addict* **7**, 252–259 (2018).

15. Open letter by scientists. Screen time guidelines need to be built on evidence, not hype. *Guardian* (2017).

16. Baron, K. G., Abbott, S., Jao, N., Manalo, N. & Mullen, R. Orthosomnia: Are Some Patients Taking the Quantified Self Too Far? *Journal of Clinical Sleep Medicine* **13**, 351–354 (2017).

17. Yin, H. H. & Knowlton, B. J. The role of the basal ganglia in habit formation. *Nat Rev Neurosci* **7**, 464–476 (2006).

18. Wood, W. & Rünger, D. Psychology of Habit. *Annual Review of Psychology* **67**, 289–314 (2016).

19. Heitmayer, M. & Lahlou, S. Why are smartphones disruptive? An empirical study of smartphone use in real-life contexts. *Comput. Hum. Behav.* **116**, 106637 (2021).

20. Macmillan, M. B. A wonderful journey through skull and brains: the travels of Mr. Gage's tamping iron. *Brain Cogn* **5**, 67–107 (1986).

21. Macmillan, M. & Lena, M. L. Rehabilitating Phineas Gage. *Neuropsychol Rehabil* **20**, 641–658 (2010).

22. Cristofori, I., Cohen-Zimerman, S. & Grafman, J. Executive functions. *Handb Clin Neurol* **163**, 197–219 (2019).

23. Guarana, C. L., Ryu, J. W., O'Boyle, E. H., Lee, J. & Barnes, C. M. Sleep and self-control: A systematic review and meta-analysis. *Sleep Medicine Reviews* **59**, 101514 (2021).

24. Arora, T., *et al.* A systematic review and meta-analysis to assess the relationship between sleep duration/quality, mental toughness and resilience amongst healthy individuals. *Sleep Medicine Reviews* **62**, 101593 (2022).

25. Blain, B., Hollard, G. & Pessiglione, M. Neural mechanisms underlying the impact of daylong cognitive work on economic decisions. *Proceedings of the National Academy of Sciences* **113**, 6967–6972 (2016).

26. Vohs, K. D., *et al.* A Multisite Preregistered Paradigmatic Test of the Ego-Depletion Effect. *Psychol Sci* **32**, 1566–1581 (2021).

27. Wiehler, A., Branzoli, F., Adanyeguh, I., Mochel, F. & Pessiglione, M. A neuro-metabolic account of why daylong cognitive work alters the control of economic decisions. *Curr Biol* **32**, 3564–3575.e3565 (2022).

28. Galla, B. M. & Duckworth, A. L. More than resisting temptation: Beneficial habits mediate the relationship between self-control and positive life outcomes. *J Pers Soc Psychol* **109**, 508–525 (2015).

29. Neal, D. T., Wood, W., Labrecque, J. S. & Lally, P. How do habits guide behavior? Perceived and actual triggers of habits in daily life. *Journal of Experimental Social Psychology* **48**, 492–498 (2012).

30. Neal, D. T., Wood, W., Wu, M. & Kurlander, D. The pull of the past: when do habits persist despite conflict with motives? *Pers Soc Psychol Bull* **37**, 1428–1437 (2011).

31. Wansink, B. & Sobal, J. Mindless Eating: The 200 Daily Food Decisions We Overlook. *Environment and Behavior* **39**, 106–123 (2007).

32. Dean, B. Instagram Demographic Statistics: How Many People Use Instagram in 2023? (Backlinko, 2023).

33. Mikulic, M. The effects of push vs. pull notifications on overall smartphone usage, frequency of usage and stress levels (Uppsala University thesis, 2016).

34. Rolls, B. J., Roe, L. S. & Meengs, J. S. The effect of large portion sizes on energy intake is sustained for 11 days. *Obesity (Silver Spring)* **15**, 1535–1543 (2007).

35. Alquist, J. L., Baumeister, R. F., Tice, D. M. & Core, T. J. What You Don't Know Can Hurt You: Uncertainty Impairs Executive Function. *Front Psychol* **11**, 576001 (2020).

36. Core, T. J., Price, M. M., Alquist, J. L., Baumeister, R. F. & Tice, D. M. Life is uncertain, eat dessert first: Uncertainty causes controlled and unemotional eaters to consume more sweets. *Appetite* **131**, 68–72 (2018).

37. Melumad, S. & Pham, M. T. The Smartphone as a Pacifying Technology. *Journal of Consumer Research* **47**, 237–255 (2020).

38. Frier, S. *No Filter* (Random House Business, 2020).

39. Manikonda, L., Hu, Y. & Kambhampati, S. Analyzing User Activities, Demographics, Social Network Structure and User-Generated Content on Instagram (2014).

40. Gallagher, B. *How to Turn Down a Billion Dollars: The Snapchat Story* (Virgin Books, 2018).

41. Wagner, K., Stories' was Instagram's smartest move yet. Can it become Facebook's next big business? *Vox* (2018).

42. Morgans, J. The Inventor of the 'Like' Button Wants You to Stop Worrying About Likes. Vol. 2022. *Vice* (2017).

43. Schultz, W. Dopamine reward prediction error coding. *Dialogues Clin Neurosci* **18**, 23–32 (2016).

44. Mosseri, A., @mosseri on Instagram. Question and Answer session on Instagram Stories. (13/01/2023).

45. Le Heron, C., *et al.* Distinct effects of apathy and dopamine on effort-based decision-making in Parkinson's disease. *Brain* **141**, 1455–1469 (2018).

46. Redgrave, P., *et al.* Goal-directed and habitual control in the basal ganglia: implications for Parkinson's disease. *Nature Reviews Neuroscience* **11**, 760–772 (2010).

47. Zhou, Q. Y. & Palmiter, R. D. Dopamine-deficient mice are severely hypoactive, adipsic, and aphagic. *Cell* **83**, 1197–1209 (1995).

48. Ceceli, A. O., Bradberry, C. W. & Goldstein, R. Z. The neurobiology of drug addiction: cross-species insights into the dysfunction and recovery of the prefrontal cortex. *Neuropsychopharmacology* **47**, 276–291 (2022).

49. Rutledge, R. B., Skandali, N., Dayan, P. & Dolan, R. J. Dopaminergic Modulation of Decision Making and Subjective Well-Being. *The Journal of Neuroscience* **35**, 9811–9822 (2015).

50. Voon, V., *et al.* Impulse control disorders and levodopa-induced dyskinesias in Parkinson's disease: an update. *The Lancet Neurology* **16**, 238–250 (2017).

51. Milkman, K. L., Minson, J. A. & Volpp, K. G. Holding the Hunger Games Hostage at the Gym: An Evaluation of Temptation Bundling. *Manage Sci* **60**, 283–299 (2014).

52. Flayelle, M., Maurage, P., Karila, L., Vögele, C. & Billieux, J. Overcoming the unitary exploration of binge-watching: A cluster analytical approach. *J Behav Addict* **8**, 586–602 (2019).

53. Moodfit Blog. Mood Versus the Days of the Week. https://www.getmoodfit.com/post/mood-versus-days-of-the-week (2022).

54. Squire, L. R. The legacy of patient H. M. for neuroscience. *Neuron* **61**, 6–9 (2009).

55. Knowlton, B. J., Mangels, J. A. & Squire, L. R. A Neostriatal Habit Learning System in Humans. *Science* **273**, 1399–1402 (1996).

56. Lehéricy, S., *et al.* Distinct basal ganglia territories are engaged in early and advanced motor sequence learning. *Proc Natl Acad Sci U S A* **102**, 12566–12571 (2005).

57. Lally, P., van Jaarsveld, C. H. M., Potts, H. W. W. & Wardle, J. How are habits formed: Modelling habit formation in the real world. *European Journal of Social Psychology* **40**, 998–1009 (2010).

58. Keller, J., *et al.* Habit formation following routine-based versus time-based cue planning: A randomized controlled trial. *Br J Health Psychol* **26**, 807–824 (2021).

59. Skipworth, W. Threads' User Engagement Plummets After Explosive Start. *Forbes* (2023).

60. Schnauber-Stockmann, A. & Naab, T. K. The process of forming a mobile media habit: results of a longitudinal study in a real-world setting. *Media Psychology* **22**, 714–742 (2019).

61. Sela, A., Rozenboim, N. & Ben-Gal, H. C. Smartphone use behavior and quality of life: What is the role of awareness? *PLOS ONE* **17**, e0260637 (2022).

62. Mark, G., Iqbal, S., Czerwinski, M. & Johns, P. Bored Mondays and focused afternoons: The rhythm of attention and online activity in the workplace. *Conference on Human Factors in Computing Systems – Proceedings* (2014).

63. Schnauber-Stockmann, A., Meier, A. & Reinecke, L. Procrastination out of Habit? The Role of Impulsive Versus Reflective Media Selection in Procrastinatory Media Use. *Media Psychology* **21**, 1–29 (2018).

64. Stothart, C., Mitchum, A. & Yehnert, C. The attentional cost of receiving a cell phone notification. *J Exp Psychol Hum Percept Perform* **41**, 893–897 (2015).

65. Mark, G. Multitasking in the Digital Age. *Synthesis Lectures on Human-Centered Informatics* **8**, 1–113 (2015).

66. Dabbish, L., Mark, G. & González, V. M. Why do I keep interrupting myself? Environment, habit and self-interruption. In *Proceedings of the SIGCHI Conference on Human Factors in Computing Systems* 3127–3130 (Association for Computing Machinery, Vancouver, BC, Canada, 2011).

67. Beard, A. Life's Work: An Interview with Maya Angelou. In *Harvard Business Review* (2013).

68. Ophir, E., Nass, C. & Wagner, A. D. Cognitive control in media multitaskers. *Proceedings of the National Academy of Sciences* **106**, 15583–15587 (2009).

69. Sanbonmatsu, D. M., Strayer, D. L., Medeiros-Ward, N. & Watson, J. M. Who multi-tasks and why? Multi-tasking ability, perceived multi-tasking ability, impulsivity, and sensation seeking. *PLoS One* **8**, e54402 (2013).

70. Raichle, M. E. The brain's default mode network. *Annu Rev Neurosci* **38**, 433–447 (2015).

71. Mark, G., Iqbal, S., Czerwinski, M. & Johns, P. Focused, Aroused, but so Distractible: Temporal Perspectives on Multitasking and Communications. In *Proceedings of the 18th ACM Conference on Computer Supported Cooperative Work & Social Computing* 903–916 (Association for Computing Machinery, Vancouver, BC, Canada, 2015).

72. Ward, A. F., Duke, K., Gneezy, A. & Bos, M. W. Brain Drain: The Mere Presence of One's Own Smartphone Reduces Available Cognitive Capacity. *Journal of the Association for Consumer Research* **2**, 140–154 (2017).

73. Mark, G., Voida, S. & Cardello, A. 'A pace not dictated by electrons': an empirical study of work without email. In *Proceedings of the SIGCHI Conference on Human Factors in Computing Systems* 555–564 (Association for Computing Machinery, Austin, Texas, USA, 2012).

74. How has the pandemic changed working lives? *The Economist* (2020).

75. Fitz, N. S., *et al.* Batching smartphone notifications can improve well-being. *Comput. Hum. Behav.* **101**, 84–94 (2019).

76. Foster, R. G. There is no mystery to sleep. *Psych J* **7**, 206–208 (2018).

77. Xie, L., *et al.* Sleep Drives Metabolite Clearance from the Adult Brain. *Science (New York, N.Y.)* **342**, 373–377 (2013).

78. Tempesta, D., Salfi, F., De Gennaro, L. & Ferrara, M. The impact of five nights of sleep restriction on emotional reactivity. *Journal of Sleep Research* **29**, e13022 (2020).

79. Scott, A. J., Webb, T. L., Martyn-St James, M., Rowse, G. & Weich, S. Improving sleep quality leads to better mental health: A meta-analysis of randomised controlled trials. *Sleep Medicine Reviews* **60**, 101556 (2021).

80. Kalmbach, D. A., *et al.* Genetic Basis of Chronotype in Humans: Insights From Three Landmark GWAS. *Sleep* **40** (2017).

81. Roenneberg, T., *et al.* Epidemiology of the human circadian clock. *Sleep Med Rev* **11**, 429–438 (2007).

82. Foster, R. *Life Time: The New Science of the Body Clock, and How It Can Revolutionize Your Sleep and Health* (Penguin Life, 2022).

83. Chang, A. M., Aeschbach, D., Duffy, J. F. & Czeisler, C. A. Evening use of light-emitting eReaders negatively affects sleep, circadian timing, and next-morning alertness. *Proc Natl Acad Sci U S A* **112**, 1232–1237 (2015).

84. Wood, B., Rea, M. S., Plitnick, B. & Figueiro, M. G. Light level and duration of exposure determine the impact of self-luminous tablets on melatonin suppression. *Appl Ergon* **44**, 237–240 (2013).

85. Tähkämö, L., Partonen, T. & Pesonen, A. K. Systematic review of light exposure impact on human circadian rhythm. *Chronobiol Int* **36**, 151–170 (2019).

86. Heath, M. A., *et al.* Does one hour of bright or short-wavelength filtered tablet screenlight have a meaningful effect on adolescents' pre-bedtime alertness, sleep, and daytime functioning? *Chronobiology International* **31**, 496–505 (2014).

87. Bigalke, J. A., Greenlund, I. M., Nicevski, J. R. & Carter, J. R. Effect of evening blue light blocking glasses on subjective and objective sleep in healthy adults: A randomized control trial. *Sleep Health* **7**, 485–490 (2021).

88. Shechter, A., Kim, E. W., St-Onge, M. P. & Westwood, A. J. Blocking nocturnal blue light for insomnia: A randomized controlled trial. *J Psychiatr Res* **96**, 196–202 (2018).

89. Hester, L., *et al.* Evening wear of blue-blocking glasses for sleep and mood disorders: a systematic review. *Chronobiol Int* **38**, 1375–1383 (2021).

90. de la Iglesia, H. O., *et al.* Access to Electric Light Is Associated with Shorter Sleep Duration in a Traditionally Hunter-Gatherer Community. *J Biol Rhythms* **30**, 342–350 (2015).

91. Grønli, J., *et al.* Reading from an iPad or from a book in bed: the impact on human sleep. A randomized controlled crossover trial. *Sleep Med* **21**, 86–92 (2016).

92. Brautsch, L. A., *et al.* Digital media use and sleep in late adolescence and young adulthood: A systematic review. *Sleep Med Rev* **68**, 101742 (2023).

93. Kroese, F. M., De Ridder, D. T., Evers, C. & Adriaanse, M. A. Bedtime procrastination: introducing a new area of procrastination. *Front Psychol* **5**, 611 (2014).

94. Hill, V. M., Rebar, A. L., Ferguson, S. A., Shriane, A. E. & Vincent, G. E. Go to bed! A systematic review and meta-analysis of bedtime procrastination correlates and sleep outcomes. *Sleep Med Rev* **66**, 101697 (2022).

95. Exelmans, L. & Van den Bulck, J. 'Glued to the Tube': The Interplay Between Self-Control, Evening Television Viewing, and Bedtime Procrastination. *Communication Research* **48**, 594–616 (2021).

96. Chung, S. J., An, H. & Suh, S. What do people do before going to bed? A study of bedtime procrastination using time use surveys. *Sleep* **43** (2019).

97. Liu, H., Ji, Y. & Dust, S. B. 'Fully recharged' evenings? The effect of evening cyber leisure on next-day vitality and performance through sleep quantity and quality, bedtime procrastination, and psychological detachment, and the moderating role of mindfulness. *J Appl Psychol* **106**, 990–1006 (2021).

98. Stothard, E. R., *et al.* Circadian Entrainment to the Natural Light-Dark Cycle across Seasons and the Weekend. *Curr Biol* **27**, 508–513 (2017).

99. Grosser, L., Knayfati, S., Yates, C., Dorrian, J. & Banks, S. Cortisol and shiftwork: A scoping review. *Sleep Med Rev* **64**, 101581 (2022).

100. Burns, A. C., *et al.* Time spent in outdoor light is associated with mood, sleep, and circadian rhythm-related outcomes: A cross-sectional and longitudinal study in over 400,000 UK Biobank participants. *J Affect Disord* **295**, 347–352 (2021).

101. Huang, L., *et al.* A Visual Circuit Related to Habenula Underlies the Antidepressive Effects of Light Therapy. *Neuron* **102**, 128–142.e128 (2019).

102. Gardiner, C., *et al.* The effect of caffeine on subsequent sleep: A systematic review and meta-analysis. *Sleep Medicine Reviews* **69**, 101764 (2023).

103. Le Beau Lucches, E. What Is Smartphone Addiction and Is It Fueling Mental Health Problems? In *Discover Magazine* (2023).

104. Walton, A. G. Phone Addiction Is Real – And So Are Its Mental Health Risks. In *Forbes* (2017).

105. Pearson, A. Highly addictive smartphones are destroying teenagers – we need to ban them now. In *The Telegraph* (2022).

106. Beyens, I., Pouwels, J. L., van Driel, I. I., Keijsers, L. & Valkenburg, P. M. Social Media Use and Adolescents' Well-Being: Developing a Typology of Person-Specific Effect Patterns. *Communication Research*, 00936502211038196 (2021).

107. House, A. Social media, self-harm and suicide. *BJPsych Bull* **44**, 131–133 (2020).

108. Susi, K., Glover-Ford, F., Stewart, A., Knowles Bevis, R. & Hawton, K. Research Review: Viewing self-harm images on the internet and social media platforms: systematic review of the impact and associated psychological mechanisms. *Journal of Child Psychology and Psychiatry* **n/a** (2023).

109. Giumetti, G. W. & Kowalski, R. M. Cyberbullying via social media and well-being. *Curr Opin Psychol* **45**, 101314 (2022).

110. Klüver, H. & Bucy, P. C. 'Psychic blindness' and other symptoms following bilateral temporal lobectomy in Rhesus monkeys. *American Journal of Physiology* **119**, 352–353 (1937).

111. Domínguez-Borràs, J. & Vuilleumier, P. Amygdala function in

emotion, cognition, and behavior. *Handb Clin Neurol* **187**, 359–380 (2022).

112. Shin, L. M. & Liberzon, I. The Neurocircuitry of Fear, Stress, and Anxiety Disorders. *Neuropsychopharmacology* **35**, 169–191 (2010).

113. Kenwood, M. M., Kalin, N. H. & Barbas, H. The prefrontal cortex, pathological anxiety, and anxiety disorders. *Neuropsychopharmacology* **47**, 260–275 (2022).

114. Gilmore, J. H., *et al.* Longitudinal development of cortical and subcortical gray matter from birth to 2 years. *Cereb Cortex* **22**, 2478–2485 (2012).

115. Tottenham, N. & Gabard-Durnam, L. J. The developing amygdala: a student of the world and a teacher of the cortex. *Curr Opin Psychol* **17**, 55–60 (2017).

116. Wadley, G., Smith, W., Koval, P. & Gross, J. J. Digital Emotion Regulation. *Current Directions in Psychological Science* **29**, 412–418 (2020).

117. Maani, C. V., *et al.* Virtual reality pain control during burn wound debridement of combat-related burn injuries using robot-like arm mounted VR goggles. *J Trauma* **71**, S125–130 (2011).

118. Neugebauer, V. Amygdala physiology in pain. *Handb Behav Neurosci* **26**, 101–113 (2020).

119. Killingsworth, M. A. & Gilbert, D. T. A wandering mind is an unhappy mind. *Science* **330**, 932 (2010).

120. Tang, Y. Y., Hölzel, B. K. & Posner, M. I. The neuroscience of mindfulness meditation. *Nat Rev Neurosci* **16**, 213–225 (2015).

121. Walsh, E. I., Smith, L., Northey, J., Rattray, B. & Cherbuin, N. Towards an understanding of the physical activity-BDNF-cognition triumvirate: A review of associations and dosage. *Ageing Research Reviews* **60**, 101044 (2020).

122. Sudimac, S., Sale, V. & Kühn, S. How nature nurtures: Amygdala activity decreases as the result of a one-hour walk in nature. *Mol Psychiatry* **27**, 4446–4452 (2022).

123. Russell, G. & Lightman, S. The human stress response. *Nature Reviews Endocrinology* **15**, 525-534 (2019).

124. Lupien, S. J., McEwen, B. S., Gunnar, M. R. & Heim, C. Effects of stress throughout the lifespan on the brain, behaviour and cognition. *Nature Reviews Neuroscience* **10**, 434–445 (2009).

125. Roozendaal, B., McEwen, B. S. & Chattarji, S. Stress, memory and the amygdala. *Nature Reviews Neuroscience* **10**, 423–433 (2009).

126. Coyne, S. M., Rogers, A. A., Zurcher, J. D., Stockdale, L. & Booth, M. Does time spent using social media impact mental health?: An eight year longitudinal study. *Computers in Human Behavior* **104**, 106160 (2020).

127. Ritchie, S. Don't panic about social media harming your child's mental health – the evidence is weak. In *i Newspaper* (2023).

128. Orben, A. & Przybylski, A. K. The association between adolescent well-

being and digital technology use. *Nature Human Behaviour* 3, 173–182 (2019).

129. Vuorre, M. & Przybylski, A. K. Estimating the association between Facebook adoption and well-being in 72 countries. *Royal Society Open Science* 10, 221451 (2023).

130. Zubieta, J. K., *et al.* Placebo effects mediated by endogenous opioid activity on mu-opioid receptors. *J Neurosci* 25, 7754–7762 (2005).

131. Wood, F. A., *et al.* N-of-1 Trial of a Statin, Placebo, or No Treatment to Assess Side Effects. *New England Journal of Medicine* 383, 2182–2184 (2020).

132. Shaw, H., *et al.* Quantifying smartphone 'use': Choice of measurement impacts relationships between 'usage' and health. *Technology, Mind, and Behavior* 1, No Pagination Specified (2020).

133. Patrick, V. M. & Hagtvedt, H. 'I Don't' versus 'I Can't': When Empowered Refusal Motivates Goal-Directed Behavior. *Journal of Consumer Research* 39, 371–381 (2011).

134. Arguinchona, J. H. & Tadi, P. *Neuroanatomy, Reticular Activating System* (StatPearls Publishing, Treasure Island (FL), 2022).

135. Costafreda, S. G., Brammer, M. J., David, A. S. & Fu, C. H. Predictors of amygdala activation during the processing of emotional stimuli: a meta-analysis of 385 PET and fMRI studies. *Brain Res Rev* 58, 57–70 (2008).

136. Hakamata, Y., *et al.* Implicit and explicit emotional memory recall in anxiety and depression: Role of basolateral amygdala and cortisol-norepinephrine interaction. *Psychoneuroendocrinology* 136, 105598 (2022).

137. Phillips, W. J., Hine, D. W. & Thorsteinsson, E. B. Implicit cognition and depression: A meta-analysis. *Clinical Psychology Review* 30, 691–709 (2010).

138. Mitte, K. Memory bias for threatening information in anxiety and anxiety disorders: a meta-analytic review. *Psychol Bull* 134, 886–911 (2008).

139. Vandenbosch, L., Fardouly, J. & Tiggemann, M. Social media and body image: Recent trends and future directions. *Curr Opin Psychol* 45, 101289 (2022).

140. Tiggemann, M. & Anderberg, I. Social media is not real: The effect of 'Instagram vs reality' images on women's social comparison and body image. *New Media & Society* 22, 2183–2199 (2019).

141. Tiggemann, M. Digital modification and body image on social media: Disclaimer labels, captions, hashtags, and comments. *Body Image* 41, 172–180 (2022).

142. Paxton, S. J., McLean, S. A. & Rodgers, R. F. 'My critical filter buffers your app filter': Social media literacy as a protective factor for body image. *Body Image* 40, 158–164 (2022).

143. Dubois, E. & Blank, G. The echo chamber is overstated: the moderating

effect of political interest and diverse media. *Information, Communication & Society* **21**, 729–745 (2018).

144. Huang, C. Time Spent on Social Network Sites and Psychological Well-Being: A Meta-Analysis. *Cyberpsychol Behav Soc Netw* **20**, 346–354 (2017).

145. Saiphoo, A. N., Dahoah Halevi, L. & Vahedi, Z. Social networking site use and self-esteem: A meta-analytic review. *Personality and Individual Differences* **153**, 109639 (2020).

146. Valkenburg, P. M., Pouwels, J. L., Beyens, I., van Driel, I. I. & Keijsers, L. Adolescents' social media experiences and their self-esteem: A person-specific susceptibility perspective. *Technology, Mind, and Behavior* **2**, No Pagination Specified (2021).

147. Meier, A. & Johnson, B. K. Social comparison and envy on social media: A critical review. *Current Opinion in Psychology* **45**, 101302 (2022).

148. Myers, T. A. & Crowther, J. H. Social comparison as a predictor of body dissatisfaction: A meta-analytic review. *J Abnorm Psychol* **118**, 683–698 (2009).

149. Hall, J. A., Xing, C., Ross, E. M. & Johnson, R. M. Experimentally manipulating social media abstinence: results of a four-week diary study. *Media Psychology* **24**, 259–275 (2021).

150. Radtke, T., Apel, T., Schenkel, K., Keller, J. & von Lindern, E. Digital detox: An effective solution in the smartphone era? A systematic literature review. *Mobile Media & Communication* **10**, 190–215 (2022).

151. Vally, Z. & D'Souza, C. G. Abstinence from social media use, subjective well-being, stress, and loneliness. *Perspect Psychiatr Care* **55**, 752–759 (2019).

152. Allcott, H., Braghieri, L., Eichmeyer, S. & Gentzkow, M. The Welfare Effects of Social Media. *American Economic Review* **110**, 629–676 (2020).

153. Reinecke, L., Gilbert, A. & Eden, A. Self-regulation as a key boundary condition in the relationship between social media use and well-being. *Curr Opin Psychol* **45**, 101296 (2022).

154. Valkenburg, P. M., van Driel, I. I. & Beyens, I. The associations of active and passive social media use with well-being: A critical scoping review. *New Media & Society* **24**, 530–549 (2022).

155. Burrow, A. L. & Rainone, N. How many likes did I get?: Purpose moderates links between positive social media feedback and self-esteem. *Journal of Experimental Social Psychology* **69**, 232–236 (2017).

156. Gilovich, T. & Kumar, A. Chapter Four – We'll Always Have Paris: The Hedonic Payoff from Experiential and Material Investments. In *Advances in Experimental Social Psychology*, Vol. 51 (eds Olson, J. M. & Zanna, M. P.) 147–187 (Academic Press, 2015).

157. Kang, S. H. K. Spaced Repetition Promotes Efficient and Effective Learning: Policy Implications for Instruction. *Policy Insights from the Behavioral and Brain Sciences* **3**, 12–19 (2016).

158. Plato. *Phaedrus*.

159. Katzman, R., *et al*. Clinical, pathological, and neurochemical changes in dementia: a subgroup with preserved mental status and numerous neocortical plaques. *Ann Neurol* **23**, 138–144 (1988).

160. Stern, Y. & Barulli, D. Cognitive reserve. *Handb Clin Neurol* **167**, 181–190 (2019).

161. Hedman, A. M., van Haren, N. E., Schnack, H. G., Kahn, R. S. & Hulshoff Pol, H. E. Human brain changes across the life span: a review of 56 longitudinal magnetic resonance imaging studies. *Hum Brain Mapp* **33**, 1987–2002 (2012).

162. Maguire, E. A., *et al*. Navigation-related structural change in the hippocampi of taxi drivers. *Proceedings of the National Academy of Sciences* **97**, 4398–4403 (2000).

163. Peng, S., Roth, A. R., Apostolova, L. G., Saykin, A. J. & Perry, B. L. Cognitively stimulating environments and cognitive reserve: the case of personal social networks. *Neurobiology of Aging* **112**, 197–203 (2022).

164. Cotten, S. R., Schuster, A. M. & Seifert, A. Social media use and well-being among older adults. *Curr Opin Psychol* **45**, 101293 (2022).

165. Khoo, S. S. & Yang, H. Social media use improves executive functions in middle-aged and older adults: A structural equation modeling analysis. *Computers in Human Behavior* **111**, 106388 (2020).

166. Balderston, N. L., Schultz, D. H. & Helmstetter, F. J. The Effect of Threat on Novelty Evoked Amygdala Responses. *PLOS ONE* **8**, e63220 (2013).

167. Szalavitz, M. Do Family Dinners Really Reduce Teen Drug Use? In *Time* (2012).

168. Miller, D. P., Waldfogel, J. & Han, W. J. Family meals and child academic and behavioral outcomes. *Child Dev* **83**, 2104–2120 (2012).

169. Musick, K. & Meier, A. Assessing Causality and Persistence in Associations Between Family Dinners and Adolescent Well-Being. *Journal of Marriage and Family* **74**, 476–493 (2012).

170. Dempsey, S., Lyons, S. & McCoy, S. Later is better: mobile phone ownership and child academic development, evidence from a longitudinal study. *Economics of Innovation and New Technology* **28**, 798–815 (2019).

171. Burrow, A. L. & Spreng, R. N. Waiting with purpose: A reliable but small association between purpose in life and impulsivity. *Pers Individ Dif* **90**, 187–189 (2016).

Acknowledgements

It's been a long and winding journey to get to this point. Years have been poured into researching, drafting, writing, erasing, redrafting, and rewriting what you now hold in your hands.

A heartfelt thank you goes out to my editor, Georgina Blackwell. True to the theme of this book, technology played a pivotal role in connecting us. Our paths first crossed as a result of my posts on social media, and we continued to collaborate through virtual meetings and email exchanges, shaping this book from its inception. Gina has exceeded every expectation I had for an editor. Her unrelenting passion for the subject, meticulous attention to detail, and steadfast patience – especially during those early, somewhat painful drafts – have been nothing short of incredible.

Thank you to everyone at Head of Zeus for enabling me to publish my work. If you've discovered my book outside of social media, it's likely thanks to the publicity efforts of Kathryn Colwell, and the marketing acumen of Jade Gwilliam and Zoe Giles.

Thank you to my long-time friend, Laily Karia, who provided not only a reliable (and much-utilised!) source of digital emotion regulation but also enthusiasm when she read the introduction, which gave me the fuel that I needed to keep going at a very critical time.

My daughters, Lyra and Aria, deserve a special mention as they have been a huge source of inspiration for this book whilst simultaneously slowing down its creation. Having children gives our brain a bigger picture, a more forward-thinking perspective. I began writing this book during my maternity leave with my second child, a sleeping baby by my side, who has now grown into an especially assertive toddler. My eldest, Lyra, would often mimic my writing process in her play, imagining herself going to a coffee shop to write her own book.

I've long been captivated by the intricate workings of the brain, and the fact that my own neural circuitry has enabled me to write this book is both humbling and fascinating. But my brain didn't construct itself in isolation – it was moulded by the influences around me, most notably my parents who, despite being no longer here, have left a lasting imprint. My dad, an avid bibliophile, often brought home bags of second-hand books that were being discarded by our local library. His love for the written word has strongly influenced me, and I find solace in the thought that if this book ever finds its way into the discount pile, someone like him might rescue it. My mum, who once tried to stuff my weighty PhD thesis into her handbag with the aim of 'subtly' showing off, would no doubt be doing the same with this book – trying to drop it into every conversation while seamlessly producing a copy she 'just happened' to have. And my grandma, a woman who lived in an era that did not allow her to pursue an education, whose stories not only enriched my life but also found their way into these pages.

Before the pivotal moment that I decided to pursue neurology, there was an even more momentous life shift. In my first year of university, I met someone who would go on to become my team mate in life. My husband, Steve, has been the foundation of this book and it is no exaggeration that I would not have been able to do this without him. There were endless discussions surrounding these topics, pushing me when my ideas did not have complete clarity. He diligently read every chapter and gave the type of criticism that only someone deeply invested in my success would. He was my source of encouragement, especially when those low power mode, self-doubting thoughts crept in. And, during this time, he shouldered an overwhelming amount of the parental responsibilities. Words will never be enough to thank him or to describe how grateful I feel for his unwavering support.

I am grateful for the many awe-inspiring individuals I've encountered on the journey to becoming a doctor and a scientist who imparted their wisdom upon me, as well as my former and current NHS colleagues. My ideas have emerged from the most unexpected places; a conversation with you might have sparked an idea that eventually found its way into this book. Thank you to everyone who has engaged with and enriched my ideas through social media. Your input has been invaluable in clarifying my ideas and honing my explanation skills. This book was made so much better by being able to visualize the people I was speaking to.

Practical Index

Index